The Way

of SCIENCE

The Way
of SCIENCE

A Philosophy of Ecology for the Layman

by
FRANK E. EGLER, Ph.D.

with an epilogue by Harry E. Van Deusen

HAFNER PUBLISHING COMPANY
New York
1970

Published by
Hafner Publishing Co., Inc.
866 Third Avenue
New York, N.Y. 10022

Copyright © 1970 by Frank E. Egler
Norfolk, Connecticut, U.S.A. 06058
Library of Congress Catalog Card Number: 79-130516
Designed by the author
Composition in Baskerville by Topel Typographic Corporation
Printed by Noble Offset, N.Y.C.
Manufactured in the United States of America

83071

to

HARRY E. VAN DEUSEN

without whose cooperating intellectual stimulation
our NINTH LEVEL

—never could have been conceived
—or born.

We are very ready
  to boast of living in a century
  of enlightenment and science.

And yet the truth is quite the reverse;
  we are still lingering
  among rudimentary and infantile forms
  of intellectual conquest.

PIERRE TEILHARD DE CHARDIN (1881-1955)
Building the Earth. 1966

PREFACE

How shall I hope to express myself to each man's humour and conceit, or to give satisfaction to all? Some understand too little, some too much, some value books by authors, as people judge of men by their clothes . . ., not regarding what, but who writes.

—ROBERT BURTON (1577-1640)
Anatomy of Melancholy

I have an impassioned faith in science, and in all that it can accomplish for our comprehension of the nature of man, of the nature of the total-environment, of the single "system" that stems from the unity that is composed of them, and not leastly of the quality that may arise from the magnificent potential of the mind of man. For this goal, modesty is mandatory. The world is not only stranger than we think. It is stranger than we *can* think.

On the other hand, let us also admit that science, like every other intellectual insight in the history of mankind, may deteriorate into the emptiness of dead dogma, into pseudo-rational irrationality, into a meaningless performance by an army of technician-privates directed by generals of like mentality, into an Establishment that ranks with church or state, with industry or the military. It may deteriorate into a mystique for the masses, to yield an imagined satisfaction of primitive cravings for the True, for the Good and the Beautiful, and even for all the Deadly Sins.

That science has been so prostituted in the past, is so prostituted in these troubled times, and will again be so prostituted in the future, that is why I have written this book, never to insult science, but in an effort to separate the blemishes from the virtues that can yet lead Man on Earth to shine supreme.

Every man, every race, every nation, every culture, every civilization has its W A Y, its way of looking upon itself, and upon its total environment, its way of interpreting them and of understanding them. Our tortured existence upon this planet, with its fleeting moments of happiness, has to "make sense." If not, suicide is committed, be it the man, the nation, or the civilization.

The Way of the twentieth century is "science," dominating our thoughts and actions whether we like it or not. It is not possible to define the word to the satisfaction of all believers. It is *the* Way of

20th century man, regardless of what he believes. Similarly, not too long ago, what we call dogmatic religion was *the* way, and people were caught up in friendly but asinine arguments, disputatious dialectics, and mammoth massacres to prove that t h e i r way was the only way. Times have not changed, only the words. Today, what man believes *is* science.

"Philosophy of science" is a recognized field of knowledge, purporting to deal with the relationships between "philosophy" (the totality of yesteryear when all knowledge was called by that name) and "science" (the totality of today when all knowledge is called by that name). Theoretically, philosophy of science is an unwanted little bastard which is not accepted into either family without being so dressed up and mannered that its hybrid parentage goes unrecognized. When taught in philosophy departments, by philosophers to philosophy students, it is a well-dressed little sprig in its family circle, speaking that dialect, and rarely, if ever, playing games with the brawny boys of other Ivied halls. When taught in science departments—and it very rarely is—it is a roly-poly brainy youth, fat with the mathematical jargon of the physicists, beloved not only by those scientists but indispensable for the maintenance of paternal sanity in their world of relativity, non-Euclidian geometries, and particle physics. For scientists in other fields, the language is not only unintelligible, but gives rise to further semantic confusions involving the words "science," "philosophy," and "religion." It is the mathematician, the astronomer, or the physicist who most often in his retirement—or his dotage (another problem of definitions)—has the God of his forefathers preside over the scientists of today. Perhaps he is right.

This book is not written for the scientists, or for the philosophers. They have their own literature. It is written for the great mass of people, reasonably intelligent people, whose lives, goals, aims, values are conditioned both by the things of science as well as by the thoughts of science. All of us need *a* Way, a way to approach ourselves, our fellow man, and the total environment about us. I firmly believe that the Way *available* today is a better Way, a Way that *can* lead to greater peace and contentment, harmony happiness and enjoyment. That same Way can also lead to destruction, not only of our civilization but of the environment also. If we are intelligent, we can choose the Way of the wise. First however, we must understand the Way of science. To this end, I write. The days of awe and of arrogance must pass. Man, as a part of nature, must understand nature, and work with nature— or he will quickly vanish from the face of the earth, a clever experiment but an unsuccessful evolutionary ephemerality.

* * * * * *

The materials and ideas in this book have developed, atrophied, hypertrophied and otherwise orbited in circles ever since I became a university professor in 1937, and especially since I tried diligently to be a drop-out from those august heights in 1945. I wish to acknowledge a very special debt of gratitude to the John Simon Guggenheim Memorial Foundation, for a Fellowship thru the years 1956-58. During that time, much material was developed as the introductory sections to a book manuscript in the field of my scientific specialty, Vegetation Science and Vegetation Management. Shortly thereafter, the book manuscript was given a class-room testing on three campuses, at the University of Rhode Island with Elmer Palmatier, at Connecticut College with William A. Niering, and at Pomona (Calif.) College with Edwin A. Phillips. To these professors and the students of their classes I am very greatly indebted. Their reactions were voluminous, candid, critical and helpful. It might be said that in some instances they were knocked off their rockers, but that once they got back on, they seemed to enjoy the ride. It was Dr. Niering who made the pointed comment that all this introductory material was really irrelevant and extraneous to the specialized science which the book aimed to survey. He urged me to delete it. His "logic" was flawless, for I know of no textbook that gives more than briefest mention to these things, or thinks it needs to. I did delete the introductory material (which did not make that book any more acceptable to ruling factions—it is still in manuscript), but I did not discard it. To the contrary, I amplified it and here present it—not for the nutrition of those with specialist-stomachs which do not produce the enzymes to digest it, and which thus pass it on for unaltered defecation, but with the hope and the faith that other stomachs have other enzymes, and that such gustatory delights may eventually produce some of the mental muscle so needed in our science and in our society.

> *It is almost as dangerous to an author as to a politician to show a sense of humor, which is to say, a capacity to discover hidden and surprising things, to penetrate to the hollowness of common assumptions, and to invent novel and arresting turns of speech. All dolts envy him this capacity, and hence dislike him. It took Mark Twain almost a generation to live down the fact that he viewed the world with laughing eyes.*
>
> —H. L. MENCKEN (1880-1956)

Drawing, page II, by LEWIS BROWNE, from *The World's Great Scriptures* by Lewis Browne, 1962, by permission of the Macmillan Co., New York, N.Y.

Text, page VI, by PIERRE TEILHARD DE CHARDIN, from *Building the Earth*, copyright 1965, reprinted by permission of the publisher, Dimension Books, Denville, New Jersey 07834.

Text, page xiv, by DESMOND MORRIS, from *The Naked Ape*, 1967, by permission of the McGraw-Hill Book Company, New York, N.Y.

Poem, page 8, by SIR WILLIAM DAMPIER, from *A History of Science* by Sir William Dampier, 1949, by permission of the Cambridge University Press, New York, N.Y.

Three figures, pages 123-125, by HARRY E. VAN DEUSEN, from *American Scientist* 52 (1), courtesy of The Society of the Sigma Xi.

Illustration, page 114, "Lucifer, King of Hell" by GUSTAVE DORÉ (1833-1883), for La Divina Commedia, by DANTE ALIGHIERI (1265-1321).

Text, page 114, GENESIS, Chapter 1, Verses 28-30.

To MARSTON BATES, my very sincere and deep appreciation. He read the entire manuscript of this book, and his ample comments and suggestions were not only helpful and corrective, but gave me a peace of mind that those who know both Bates and Egler can understand.

The Book Essays

The would-be author-scholar, to impress his peers in Academe as well as the great outside world which he hopes will buy the book, considers it de rigueur to spot his pages with impressive footnotes (that no one reads), and to conclude the volume with a lengthy bibliography (that no one uses). Now it is true that there is a kind of scholarly contribution for which this type of documentation is a reasonable and necessary adjunct to the progress and development of the author's argument. On the other hand, I have long sensed that these procedures often reflect an inner insecurity on the part of the author, and a pathologic compulsion to have a "father" to fall back upon in case of challenge. I have often seen the most obvious and accepted views and facts so "footnoted." As for the so-called bibliographies, "Literature Cited," "References," and "Recommended Readings": bibliographies on special subjects are separately published and easily available. Why reprint them? In reviewing one manuscript for an ecologist, I recall accusing him of padding his bibliography (it was almost as long as the paper itself) with everything he had ever read in the field as a sort of "See how much I have read!" Perhaps the twist is in many of us. Bibliographies would be of value IF they included an annotation and evaluation by the author, the man who has supposedly looked at them and found them significantly related to his own book. Too often I have spent time digging unannotated references out of libraries, to find them utterly trivial, or unrelated to my further interests. There seems to be a conspiracy on the part of authors, editors and publishers to forbid all such annotations as a sort of Value Judgment unworthy of an objective scientist. Such, is part of the Way of Science, which is the subject matter of this volume.

In an attempt partly to offset this mannered folkway of my tribe, I am including a few "book essays." These short essays are not "reviews." Reviews are themselves interesting expressions of personality-types. They may be cold and unvalued statements of the contents, or derogatory nit-picking on mis-spelled words and technical omissions, or polemical explosions designed to give a yapping puppy a sense of superiority over a giant whose greatness he but dimly grasps. And occasionally, very occasionally, there are reviews that live longer than the books they review. Someday I hope that psychologists will study the psychology of reviewers.

The Book Essays I have included are not "reviews." They are short essays dealing with the book from a personal and limited viewpoint, in relation to the particular section of *The Way of Science* at which they are inserted. The few that are included were chosen with indefensible arbitrariness. I liked them, for any one of many reasons. I offer my very sincere and humble apologies to many many authors, recognition of whose volumes could also grace these pages, also for any one of many reasons.

It is my intention to continue to prepare additional Book Essays, relevant to specific sections of *The Way of Science*. If any of my readers wish to receive "annual supplements" of such collections of essays, the request should be made directly to me.

List of Book Essays

List of Figures

Despite our grandiose ideas and our lofty self-conceits, we are still humble animals, subject to all the basic laws of animal behavior. We tend to suffer from a strange complacency that there is something special about us, that we can never collapse as a dominant species, that we are somehow above biologic control. But we are not. Many exciting species have become extinct in the past, and we are no exception. Sooner or later we shall go, making way for something else. If it is to be later rather than sooner, then we must take a long, hard look at ourselves as biological specimens, and gain some understanding of our limitations. This is why I have written this manuscript and why I have deliberately insulted us.

—DESMOND MORRIS
The Naked Ape. 1967

Table of Contents

Where We ARE

THE ACTUALITIES OF TODAY

THE NATURE OF SCIENCE

True science teaches us to doubt,
And in ignorance to refrain.

—CLAUDE BERNARD (1813-1878)

Mankind loves a mystery, but it hates a mystery more. Thus humanity has explained and systematized its perceptions of the external world, and called the resultant "knowledge." Knowledge is not wisdom; wisdom is knowledge, when it is tempered by judgment.

There is more than one road to knowledge. That of *superstition* is linked with witchcraft and quackery, myths mystiques and magic, fads, fancies, frauds and fakes. Mysticism, occultism, theosophy and some of the world's religions rely on *mystical insight*. Some dogmatic religions use *divine revelation* or *infallibility* as a source. *Science* is just another road to knowledge that depends on certain rules and regulations which are considered, at least by their makers, as rational and reasonable. These different roads to knowledge are not mutually exclusive. Many people accept different roads on different days of the week, or for different types of knowledge, or at different ages of their lives.

 "Knowledge is not wisdom; wisdom is knowledge, when it is tempered by judgment."

There are many definitions of science, some extremely simple and some clothed in a jargon only intelligible to the initiated. By definitions alone, it is often difficult to find a common core among the diversities. As knowledge of the world about us, we refer to the so-called "natural sciences," the function of which is to describe and explain the things, properties and phenomena of nature. To others, science is the accumulation and classification of "facts." To still others, it is the organization of these facts, by means of concepts, into a coordinated body of knowledge. From another point of view, science is a philosophy, a method, a viewpoint, an attitude, one of several that may explain the world about us. By one segment of the public, science is considered. like the medicine men of old, to be the source of wonders

1

and miracles. However, for the general public, there is simply nothing that the god science cannot accomplish, given a big enough team of men and enough money thrown into the project. By other segments of the public, science is a sacred cow, a be-haloed community of high priests their rituals and their worshipful followers. Undoubtedly science is a bit of all these things, ever reflecting a faith in the intelligibility of nature.

At best, science is a body of knowledge of us and the world about us, systematized, organized, with accepted principles and concepts. It is allegedly logical in its explanations, purportedly reasonable in its methods, hopefully reliable in its predictions—an ever-accumulating heritage from the past, presumably progressing unidirectionally from fallacy and phallicism towards a fealty to fact.

> *". . . science . . . ever reflecting*
> *a faith in the intelligibility of nature."*

Scientific knowledge is often a phenomenon of a particular time. History can either redden or pale the face of many a scientist were he to know (and he usually does not) how often the science of yesterday has become the superstition of today. By analogy, much of the science of today might be destined to become the superstition of tomorrow. The thought is humbling.

Scientific knowledge passes gradually and imperceptibly into other types of knowledge. In one direction, it fades into those pseudo-sciences which have set so many souls afoundering. In a totally different direction, it passes into the technologies and engineering fields, which by their nature, must remain predominantly empirical, rule-of-thumb and "practical." These are the applications of knowledge that have created our present industrial and economic society and that are metamorphosing the face of the earth, presumably for our greater happiness, if happiness is defined as satisfying our compulsion to make those changes.

Like every other Way, science has its taboos, sacred interdictions against deed or thought, even the sin of raising questions, taboos such as are common in all primitive cultures. Oddly, the sciences involving man himself are the last to escape the clutches of his taboos. For example, psychology had a long struggle to become respectable. Altho Charles Darwin himself wrote a book entitled *Expression of the Emotions in Man and Animals,* it is taking us a full century gingerly to accept the idea that human behavior, as well as human anatomy and human physiology, are the product of biologic evolution. The entire complex of the "social sciences"—despite the empirical formulations of the insurance industry, the prognostications of the economic advisors to government, and the role of the Military as mankind's major, most costly, most destructive activity—is looked upon by the scientists in power as an unworthy aspirant to their Ivory Tower. The mere mention of "racial differences" not only arouses no scientific question in their minds, but brings forth such emotional and dogmatic assertions of biologic "identity," "liberty," "fraternity" and "equality" as are on an intellectual par with the opposing racial prejudices of history's most venomous dictators. And within this supersensitive field, the doors of knowledge are locked tight against the unholy thought that there might be racial differences in "intelligence" (by whatever standards that moot concept may be defined and investigated), or that there might be variations in the quality of human inheritance, both for individuals, and for interbreeding populations. Indeed, I trust that some scientist of the future might make a study, not of racial differences in intelligence and quality (for that taboo is bound to be broken) but of the gyrating ratiocinations by which otherwise intelligent scientists argue convincingly (amongst themselves) that they cannot possibly prove that there are such differences. They never apply these same arguments to their accepted fields of science. If they did, the entire structure of science would fall, and with it their pride, their ego, and their jobs.

 "mention of 'racial differences' . . . arouses no scientific question."

We see that science is not just a product of man. It is a product of men. It involves the communication of knowledge between and amongst men. Science arose thru talk, thru language. And what is language?

Anthony Standen

SCIENCE IS A SACRED COW

New York: E. P. Dutton. 221 pp. 1960

Chemistry Professor Standen has written one of the very few books with a general thesis similar to my own, yet it is in a vastly different style and manner. The author is educated, literate, and widely read. In addition to introductory and concluding sections, the book comprises separate chapters on physics ("Science at its best"), Biology, Psychology, Sociology and Mathematics ("True Science").

The author is angry, scornful, derisive, opinionated, and given to sweeping absolutist statements that—no matter how justified they might be—rouse one's hackles. It is clear that science was once his own Sacred Cow, and his present blanket denunciations are a reaction to the continuing public-directed babel of its clergy and the worship of its adoring congregations. Yet it must be admitted that the man does have the clairvoyance, the clarity of vision, of the virile male who views a convention of DARters and, piercing the cosmetics and couture, the wigs brassieres and girdles, turns away in repulsion. Yet many fine and real values are thereby ignored; some gross injustices are committed.

This is a book to make scientists and science-worshippers violently angry. It is a measure however of the objectivity and impartiality of which the author speaks, to be able to read this book and to grasp the solid logical factual basis that underlies his many accusations. I would say that a scientist is not a good scientist unless he can slowly and coolly enter into this volume, and admit to the gross distortions of the scientism of which his colleagues, and probably himself, are guilty.

SCIENCE, PRODUCT OF LANGUAGE

The audible emissions of captive dolphins under water or in air are remarkably complex and varied.
                              — (From a research article by a zoologist)

As children, we learn to talk our native language. Quite literally, we imbibe it with our mother's milk. Language thus becomes for each and every one of us an unconscious, unanalyzed, unquestioned part of our existence. It is part of our ecologic environment, and part of our communication with that environment, both by the spoken and written word. We "take it for granted," even as yokels and tribesmen take for granted the air around them.

Linguistics is the science of language. It is the study of human speech in all its forms, its origins, nature, structure, modifications. It includes phonetics, syntax, grammar, semantics. The physicist is involved with the analysis of this sound only as vibrations in the atmosphere. World politics is concerned with sound as a problem of translations amongst languages. Linguistics, for all its importance, is a science in its infancy, and *comparative linguistics* is barely born.

    ". . . all higher levels of thinking
                 are conditioned by language . . ."

It is the thesis, the hypo-thesis, of my discussion that Western Science is uniquely the product of western language. This western science is a world-view, a philosophy, dictated by and dependent upon, the restrictions and limitations of a few contemporary dialects of the Indo-European tongues. It has spread throughout the world, not by virtue of being truly translated into other languages, but because its own linguistic integument has been grafted onto other tongues. Admittedly this idea is a working hypothesis. As comparative linguistics develops, and the basic natures of Amerindian, African and Oriental tongues are laid bare, new and now-unimagined relations are destined to unfold.

It is the first of two major premises that all higher levels of *thinking* are conditioned by language, even if we never speak or communicate by writing. The very concepts we employ in our creative mental processes are determined by the available words that come to mind. Unspoken language helps to shape the impressions of the outside world that come to us thru our senses. Language is a molder of ideas, an analyzer or a neglecter of nature, a guide for our mental activity, a constrainer of our abstractions, a reducer of our logic to a languid lexicon. If these ideas strike as strange, if not fanciful, then stop to realize that all your thoughts, as they are forming, are molded into terms of subject and predicate, of nouns and verbs, of the forms of those nouns in terms of gender, number and person, of the tenses of our verbs, of three-dimensional space, of the sharp cleavage between space and time (remember the problem of Einstein in overcoming the bonds of language). These are built-in arbitrary characteristics not of the world about us, but of the language in which we think, in which we can o n l y think.

> "*. . . arbitrary characteristics . . .*
> *of the language . . . in which*
> *we can o n l y think.*"

It is the second of two major premises that all higher levels of spoken or written *communication* are fettered by the chains of the language in which they are expressed. For example, archaic Chinese, and even the modern literary language, has no active and passive, no singular or plural, no case, no person, no tense, no mood. Almost any "word" can be used as almost any part of speech, and even the important part of an ideogram can be omitted, leaving what we would call a "prefix" or a "suffix." Other languages, and I think particularly of certain Amerindian and African tongues, have refinements that make the West European dialects look primitive by comparison, embracing modes of both thinking and expression that are all but untranslatable into the Indo-European. In their nouns, their verbs, and their grammar, they can convey understandings of the world with a far-reaching precision.

There is a tendency in the entire world for English to become the language of science. "Basic English" unfortunately is an eviscerated British English, with all its concealed premises exaggerated. It is being fobbed off on an unsuspecting world as the substance of pure Reason itself. Even worse, it is being made the vehicle for automated transla-tion, on the assumption of word-by-word equivalents, or at the most, of equivalent few-word groups. The limitations of archaic Chinese may be born anew, a misguided ideal enchaining the human mind, far more than ever did that ancient tongue.

> *". . . the . . . computer. . . .*
> *Undoubtedly it does think, far*
> *better than most people."*

The restrictive relation of language to science is receiving a new threat in the mid-twentieth century with the rapid ascendancy of the digital computer. This technologic construct is widely advertized as being able to "think." Undoubtedly it does think, far better than most people. On the other hand, tho fiendishly complicated and ever be-coming more so, the computer has but a small percentage of the neurons that the human brain possesses. Furthermore, it can think and act only in terms of *two* digits, zero and one, "off" and "on." Now perhaps our human brain, in its ultimate details, also thinks in this way. On the other hand, it is man which programs the computer in a 2-digit language. It is frightening enough to realize that science must suffer by the limitations of human language, even worse by English, even worse by Basic English. But when science begins to be limited by translation into a 2-digit language, and when a democratic majority will put their faith into the certainty of a technician-programmed computer, rather than the subjectivity intuition and insight of a Dar-win or an Einstein, then we have returned to the idol-worship of the dawn of civilization. I would prefer the idol, for oracles from idols did occasionally show insight.

SIR WILLIAM DAMPIER

"Natura enim non nisi parendo vincitur."

At first men try with magic charm
    To fertilize the earth,
To keep their flocks and herds from harm
    And bring new young to birth.

Then to capricious gods they turn
    To save from fire or flood;
Their smoking sacrifices burn
    On altars red with blood.

Next bold philosopher and sage
    A settled plan decree,
And prove by thought or sacred page
    What Nature ought to be.

But Nature smiles—a Sphinx-like smile—
    Watching their little day
She waits in patience for a while—
    Their plans dissolve away.

Then come those humbler men of heart
    With no completed scheme,
Content to play a modest part,
    To test, observe, and dream,

Till out of chaos come in sight
    Clear fragments of a Whole;
Man, learning Nature's ways aright,
    Obeying, can control.

The changing Pattern glows afar;
    But yet its shifting scenes
Reveal not what the Pieces are
    Nor what the Puzzle means.

And Nature smiles—still unconfessed
    The secret thought she thinks—
Inscrutable she guards unguessed
    The Riddle of the Sphinx.

Hilfield, Dorset
September, 1929

SCIENCE IN HISTORY

A new scientific truth does not triumph by convincing its opponents and making them see the light, but rather because its opponents eventually die, and a new generation grows up that is familiar with it.

—MAX PLANCK (1858-1947)
Scientific Autobiography. 1949

(As quoted by Bernard Barber in "Resistance by Scientists to a Scientific Discovery. This source of resistance has yet to be given the scrutiny accorded to religious and ideological sources." Science 134 (3479) :596-602. Sept. 1, 1961.)

It is impossible to write a history of science itself—if only because "science" cannot be rigidly defined. If it is strictly defined, the definition is arbitrary, irrational, and harmful to a very understanding of the problem it seeks to solve. Any definition is on a par with a local argot in the vast space and time of human history. Semantic problems abound at every spot in that temporal unfolding. There was always "knowledge," back before the time of recorded history, and *it* was the science of its day. For these reasons, the history of science is like a braided rope, with each strand of changing textures and colors. Philosophy and religion are intimate parts of the rope, nor can we ignore magic and witchcraft. Nor contemporary foibles.

The simple among the scientists possess a naive faith in what might be called the "Progressive Development" idea. This is a pleasant and gratifying thought, perpetuated in all the textbooks written by the platoons of specialists who are impressed by the changes, discoveries, advances and breakthroughs in the few years since they themselves finished their formal indoctrination (otherwise called education) . The idea of Progressive Development even has a broader base. Many specialists, confined in the concentration camps of their Judeo-Christian culture, mumble the ritual of unidirectional cumulative "progress" from antiquity, thru the Middle Ages, on to the glorious Renaissance, to the explosions of modern times. Yes, there is progress, of a kind.

Yet man should be more humble. Nor need we look upon the starred infinity above us to feel humility. Beneath our feet, entombed and forgotten in dust and rubble, lies a succession of knowledge-founded civilizations that struggled upwards, achieved greatness, and perished before us. They achieved a greatness in mathematics, in astronomy, linguistics, calendrics, not to mention the arts and humanities, before which we still marvel. These were the civilizations of the Indus, the Tigris and the Euphrates, the Nile, and Middle America. Depending on how one counts, there are many more. Cultures can be destroyed by floods, vulcanism, military conquest and destruction, overpopulation, famine and civil strife, environmental pollution, or there may be uncontrollable internal decay such as is being witnessed since the end of World War II. It is time to mourn the passing of the greatest single empire the world has ever known, which once ruled the waves and upon which the sun never set, and which came closest to the ideals of a peaceful global empire.

It is not easy for one to grasp the social fragility of science and of all knowledge, the utter and complete destruction to which it may be subjected without leaving even memories and legends. I know no better example than to travel through the jungles of Middle America. There are places where all day one will pass an endless series of the house platforms of the Mayan populations, but almost never a human being. And then one will camp for the night with a couple and their one child (all others had died), their hearth fire in a hollow amongst the stones of a mighty foundation, their corn patch in the opening of a windfallen jungle tree. Ask them of the history about them. The things are only the "ruins," a word used with no more depth, curiosity or understanding than the "river" which flows nearby. Nor need we wait two or three centuries for this to happen. Only two or three decades may be enough. In southern Mexico during World War I, an extensive hemp industry developed at certain places; towns for the agricultural laborers were built, and pretentious manor houses by the entrepreneurial bosses. I recall seeing one such mansion during the days of World War II; the shell remained, no windows, but an essentially intact roof. Beside the still serviceable fireplace in one of the empty rooms a couple were living, with their few possessions on the floor. What was this place to them? A "house," as others would say a "cave." They could say no more. They knew no more.

Even if we limited our perspective to our own single culture, we really find no orderly progressive development. There are meanderings, divergences, set-backs, periods of advance followed by declines. True, great individuals stand out, from Darwin and Galileo thru a host of others. Yet, if we look closely at these individuals, we find they also embrace thoughts and ideas that would horrify a later age. No one man can do more than advance a few steps before the front ranks of that contemporary army of thought to which, whether he would or not, he belongs. Advance too far too fast, and he is ignored, or killed. For the same reasons, what are in later years recognized as rank superstitions and myths tend to persist in ways that now surprise us. In 1857, for example, a well-informed individual seriously contended that God had put misleading fossils into the rocks, to test the faith of mankind. In 1967, we were oscillating as to whether quasars are the most distant objects ever seen by man, occupying the outer limits of the universe itself, or whether one of them is figuratively in bed with us, closer than nearby galaxies.

If one does wish to generalize about the progress of science, one might propose three paradoxical tenets. The first is that we progress from greater knowledge to lesser knowledge. It is the ignorant however who think they know the most. It is the youngest professor who is most cocksure that he has all the answers (unless indeed the professor fossilizes at this stage of youth). The second tenet, related to the first, is that we progress from certainty to uncertainty. The fundamental religions of old were the most "certain." Even if the Quest for Certainty is a major psychologic compulsion of science, the attainment of that certainty means its destruction, for science can progress only by change. And thirdly, progress is attained as science moves from nature-in-the-human-image to an idea of nature-in-non-human-images. The earliest sciences were sheer animism, the attribution of human spirits to rocks, trees and animals, and eventually to a powerful and almighty universe. Actually man creates nature and his gods in his own image. It is, however, more satisfying to say the obverse, that god creates man in his image, an idea that is always the last hold-out of non-science in a civilization.

And now we limit our view to the body of contemporary science; and again we find whirlpools and backwaters, and dissenting currents. The political ideologies of nations are themselves powerful scientific forces. In the 20th century, we can point to the "master race" anthropology of Nazi Germany, and the Lysenkoist agriculture of Stalin's Russia, both of which deserve the careful study of historians. Within general populaces, we can find the crassest ignorances and superstitions. Astrology has a perennial fascination for the human mind. Our newspapers and magazines continue to satisfy those cravings with endless horoscopes, and with the literature of the societies of their devotees. Who can say that the stars do not really guide our destinies when corporation officials and leading politicians are known to make their decisions by obeisance at these altars?

We cannot state that non-science is limited to the ignorant and the uneducated. The vast and ramifying body of knowledge is so fluid, so changing, so overwhelming that even some of the most intelligent and educated throw up their hands in utter discouragement, and revert to some simple dogma that is certain and satisfying. Thus we have numerous and powerful groups oriented towards two of man's focal interests, food and health. The food faddists have proliferated to unbelievably non-scientific pastimes. Perhaps the most "simplifying" is promulgated amongst the "educated," where a little knowledge is most harmful. One of the strangest such groups are those who have decided that all "chemicals" are "artificial," "unnatural," and "bad," and that only "organically grown" food is "good." Clearly, their elementary chemistry is non-existent. Parallelly, the health faddists have gone to wild and outlandish extremes. As with the food faddists, open fraudulence and medical quackery are hard to separate from well-intentioned even if ignorant simplicity. Here also, the complexity of medical knowledge is swept away, and replaced by panaceas, or cure-alls, or simply by denying the existence of ill health. Such philosophies are both unscientific and un-Christian—except to those who follow them.

I WOULD AGREE with those whose opinion it is that the vast and imposing structure of modern science is probably the greatest intellectual, cumulative triumph of the human mind throughout its long history of evolution. But this is no firm, stable, graspable, preservable "body." On the one hand, its stability is a probability. The arbitrary and coincidental discovery of new facts may at any day result in a kaleidoscopic change of fundamental and basic concepts and ideas. On the other hand, the balanced, dispassionate scientific mind is simply repellent to the many who cannot hold their judgment in abeyance while there is no valid evidence on which a judgment, any judgment, can be formed. Disagree with some person on a stand, a reason he has taken, a cause he has decided upon, be he either educated or uneducated, and he will probably angrily reply, "What *else* could it be?" These dangers to knowledge and to science grow and dominate whenever and wherever man becomes more swayed by emotion than by

 ". . . man has but one run of history from which to learn, and there will be but one run in the future."

reason, and whenever and wherever our individual and basic needs for such as food, cover, sex and space are not adequately met—even if because of our own faults. The continuance of scientific knowledge thru history, and all that is dependent upon it, demands our constant and active endeavours. If we cannot learn from history, we are fated to repeat history, in all its gory and tragic mistakes. In other words, the wise man learns from the experience of others, the fool only from his own. Unfortunately man has but one run of history from which to learn, and there will be but one run in the future.

NATURE OF THE SCIENTIST

*"And now appears a great fact of human brotherhood—that human beings
are all alike in this respect: So far as we can judge from the systematics of
language, the higher mind or 'unconscious' of a Papuan headhunter can
mathematize quite as well as that of Einstein; and conversely, scientist and
yokel, scholar and tribesman, all use their personal consciousness in the
same dim-witted sort of way, and get into similar kinds of logical impasse.
They are as unaware of the beautiful and inexorable systems that control
them as a cowherd is of cosmic rays. Their understanding of the processes
involved in their talk and ratiocination is a purely superficial, pragmatic
one . . ."*

—BENJAMIN LEE WHORF
Language, Thought and Reality. 1956

In the civilization of the mid-20th century, the growing body of
those who call themselves scientists constitute the reigning hierarchy,
the ruling priesthood, even the military might. This amazing phenomenon was triggered in 1945 by the explosions over Nagasaki and
Hiroshima. Such power! It grew by leaps and bounds until in 1964
it was estimated that six million souls around the world belonged to
this chosen minority. There are parallels everywhere with the older
institutions of social organizations. Neophytes feel "the call" (prose-
lytizing in the schools). One takes holy orders (in the halls of academe). There is ordination (the equinoctial June rites, with a parch-
ment still mainly in Latin, with the officiating priests still in the
vestments of medieval cap and gown.) There is priestly installation
(one's first job). Then "tenure" is sought (dissolved only by excom-
munication), and a lifetime of "publish or perish." If one does not
perish, there are the ritualized periodic councils (meetings, conferences, symposia). The clergy contacts the laity thru preachment to
vast congregations (the lecture system), with occasional observances
of confessional sacraments (the examination system). The lay brethren
who have sold themselves for a fast buck apply all these principles to
the economic life (technology is the foundation of our corporations).
And the Crusades are restaged by the Military, to the tune of "Onward
Christian Soldiers," while the different sects compete in their races of
lunacy and venery to reach the moon and Venus. Need I go on?

We are led to look upon the scientist as a member of a very special
breed, one "chosen" by some higher power in the universe to lead the
destinies of the human race. But scientists are people, and the scientist
is but a member of the human race, a fact frequently forgotten, but
easily verifiable by experiment. It is well if we focus upon the scientist
a bit more closely.

The scientist is believed to have a formidably high level of intelligence. Indeed it is true that all data from intelligence tests show that
scientists have an average IQ far above the levels for the general popu-

lation. I have seen no comparable data however for analogous segments from religion, industry or the military, or even professional athletics. It is further to be remembered that these tests are made by the scientists themselves, and the parameters are those that they know and value. I recall the study made by some brotherhood of barbers upon baldness (obviously a factor of economic importance in their profession). They found a striking correlation between baldness and virility. I did not read of their objective measurement for virility; but it was stated that a large majority of the researchers were bald.

With the mid-century growth of the army of scientists, educators have estimated that the high IQ sector of the population than can profitably be trained as scientists will have been fully tapped by the ¾ mark of the century. Some of us believe this has happened already. When and as it may be so, it follows that "the average man" can become a scientist. Then to be a scientist is a matter of "training," not a matter of inherent superiority even if we cannot precisely define the factors that constitute scientific superiority. This situation of course is entirely right and justifiable in the light of the political ideology that all men are created equal, and have the right to become whatever they wish in that society. This is the democracy that followed Pericles— and soon collapsed.

I do believe that the scientist is, or should be, a superior citizen. Yet his objectives are involved with what is possibly the grandest intellectual creation of the human race. He makes the noblest and sincerest attempt to get beyond the limitations of our physical senses, to conceptualize beyond the narrowness of our brains, and to understand the nature of the universe of which we are but infinitesimal mites. A scientist is a mighty mite, an intellectual. He is a lover of the intellectual. But all love is an emotional phenomenon. A scientist, with other men, is the only animal that blushes—or needs to, as Mark Twain once remarked.

Scientists are only men, and are subject to all the foibles of their kind. They have the same drives for freedom, security, certainty, image and status as have other men. They have the same attraction for the known the familiar and the comfortable, and will cling to old and sterile ideas like a broody hen sitting on boiled eggs. Like those others, there is a lunatic fringe, and a reasonable quota of social misfits, small-pool big-frogs, megalomaniacs, prima donnas, nymphomaniacs, gold diggers, entrepreneurs, prophets and devout disciples. It is significant, however, that these appelations are taken from other fields, not from science. Like other human beings, scientists are excellent at rationalizing their behavior. Indeed, they are superior, for their rationalizations are couched in academese and scientese which fools others as well as themselves. I confess very candidly that scientist-behavior had me puzzled for years and years and years—until I became personally familiar with certain laymen, with the ignorant, the uneducated, with paranoid schizophrenics, hypochondriacs and other obsessive and compulsive types of interest to the psychiatrist. Then the veil before my eyes disappeared.

If there is any one trait to characterize the leading scientists, I would say it is in the possession of an inordinate amount of curiosity, combined with enough superior intelligence to apply the rather exacting rules of the scientific game. Curiosity alone does not make a scientist—consider children, and morons—but it is highly important. A budding scientist without this bump of curiosity is likely to blight in the bud. This curiosity, this compulsion, to discover, takes several dominant forms. The scientist asks "what?" and gives us a description. He asks "How?" and gives us an explanation. "Why?" and gives us a cause. "Where?", a place. "When?", a time. We shall see later that these humanly oriented questions are responsible for the structure of our science (not the structure of reality.)

Lest the reader think otherwise, may I assure him that I count many scientists among my friends, people for whom I have respect, admiration, and regard, people whom I like. (Perhaps I should add that their qualifications as scientists are entirely irrelevant to this friendship.)

Science interdigitates with society in a multitude of manners. There is one social sequence in which scientists play a dominant role, and which is here worthy of discussion. I refer to the sequence from the Science Teacher to the Science Researcher, from the generalist to the specialist, from the kindergarten to the post-Ph.D. level. It is a continuum of its kind, with a remarkable gradient of personality types, involving emotional educational and intellectual factors. There is also unfortunately, a highly unjustified snobbery of each stage towards the stage just below it. Two stages below, and the continuum is not even recognized.

Kindergarten to Sixth Grade: The K-6 level is so generalized in its educational nature that scientists per se are not employed as teachers. As a group, these people are remarkably friendly; they love children and are loved by them. They can meet children on their own level, yet are still able to guide and impart knowledge to them. I include the group in this discussion, for in my opinion it holds the greatest single potential for educating the nation in the ways of sound scientific thought. Religion has always known this ("Get them while they are young") ; science has yet to learn it. It is my firm conviction that even the higher intellectual concepts of a Total Ecology, of conservation and of natural resource management, of human overpopulation problems, can be taught at this level better than at any other level.

Seventh-Eighth Grades: In our present educational system, these two years are betwixt and between. These sexually maturing people are past the time of Primary education, and yet not ready for the Secondary schools as we now have them organized. I know of no scientific programs at these levels worthy of comment.

Grades Nine to Twelve: These are the four years of our high school systems, composed of a good wild active group with a premature sense of their own maturity. The teaching force is probably the single most powerful educational influence in our nation. As a group—I have been with these people for many years in teacher-training programs—I have the very highest regard for them. They are generalists in mind. (They could not possibly be specialists, for as science teachers, their charges expect them to "know everything.") They have their hands on the pulse of other aspects of society, on parents, on the community, on family and social problems. Theirs is no captive audience; they m u s t be liked by the students. Consequently, by personality, the high school science teacher tends to be a friendly extroverted intelligent and educated type—in comparison to other parts of the continuum. Altho I have been told that the high school student is already too old and set-in-his-ways to be easily teachable in scientific thought (he has been already trained, by TV and the street and his own peers), and altho high school science teaching is far far from an ideal, here I believe is the greatest practical potential for upgrading the level of our citizenry.

I am tempted—with fingers crossed—to enter a special word of praise for the Catholic sisters who teach biology. I have known a reasonable number of them, and what they have in common is such a remarkable combination of desirable human traits that they alone could maintain my faith in humanity when all else boils in bitterness. They are human beings, born to teaching and born to service, with a generosity and a selflessness that would do credit to any of the founders of the world's great religions. Their humor is unsurpassable, even to such a trying time as for that unveiled virgin aboard an oceanographic vessel, when the passing chain of a winch took her black scarf to the depths below. Or when I took a 20-year-old Mother to dinner at Mory's (by special summer dispensation)—probably the first unmarried mother ever to enter the portals of Yale's famous all-male eating club, and escorted by a professor at that (1965). Untempered by the urge for ever-higher wages and fringe benefits, with what surely is a superior education and intelligence, they go thru life with a spiritual happiness and an inner harmony whereby age never seems to take its toll upon their countenance. They seem to forgive my own not-well-veiled irreligion, even while I have never known a single instance, a single innuendo, that they think me a heathen in need of a missionary. They flow thru life with an energy and grace. The energy is evident as, undaunted by flowing skirts, I have seen them climb rocky taluses of gigantic boulders like a squirrel, hop on and off hay wagons like a frisking pup, and skim across swamps and marshes like a low-flying swallow. Their grace is never lacking. I longed for a candid movie film, one day in a mid-West university cafeteria, as two females walked across an open space in full view of the diners. The one with a slouching ungainly trot, who if she fell on her face would make a 6-point landing, knobby knees, knobby stomach, knobby sweater, knobby nose. The other glided by, all smiles and flowing grace, the epitome of femininity. A bachelor, some bachelors, even many married men, might think about a wasted and unutilized natural resource. But the "multiple use" of the conservationist is a concept that can be overdone; and the fertility of renunciation is a virtue that demands our most profound obeisance.

The College Level. It had become essential previous to 1960 to separate the small liberal arts college, together with the community college, from the undergraduate years of the large university. The college had been essentially terminal in the training of the citizen. The undergraduate student of the university on the other hand was rapidly finding himself in preparatory training for a graduate specialty, whether he willed it or no. In the 1970's more and more of the college students feel compelled to continue for a graduate degree. The two faculties, even tho they have much in common, continue to have much apart. The university professor is primarily a research specialist, for whom students are just so much of a hindrance to his research career, like passenger traffic to our freight-carrying railroads. More of them later.

The college professor is a creature very much apart from the high school science teacher. He arrived where he did by virtue of being a specialist (he need have had not a single course in education, nor even any experience in it). By knowing his subject he was supposed, by that fact, to be a good pedagogue. (By contrast, the high school teacher more often than not, is taught *how* to teach even tho he does not know *what* to teach.) This matter of specialization has insidious and far reaching implications. The specialization of which he is so proud is not in "science" or even in "biology," but in some minute branch which has very little bearing on the general elementary courses he is supposed to teach. Consequently, those elementary courses are either overloaded with the professor's specialty, or they are very poorly weighted for the total subject. I believe it is true that some of the best teaching is done in some of the smaller liberal arts colleges, but in our post-Renaissance knowledge-expansion, there is always the stress for more specialization and less generalization.

In terms of temperament-type, this trend to specialization has a very insidious influence on the education of our college generation. There is a definite negative correlation between the degree of one's specialization and of his "generalization." The more of one the less of the other. Sooner or later, one's ignorance of the generalities of life starts reminding us of the idiot-savants, those strange brains who are absolute geniuses in one small field of knowledge, but dolts and morons in almost everything else. The gradation is gradual from one extreme of person to the other, but nowhere is the start more abrupt than in liberal arts faculties. I was made grimly aware of this during 1967 at a several-day conference of half-a-hundred, including college presidents, deans, and faculty members. In fact, the one man who was able to pull the conference together on the last day, and to leave it on a constructive note, was a generalist who unbeknownst to the others had had no formal education beyond high school.

> *A fool's brain digests philosophy into folly, science into superstition, and art into pedantry. Hence university education.*
>
> —G. BERNARD SHAW (1856-1950)
> *Man and Superman*

The University Level. At the next stage in this hierarchy, we have reached the highest temple of the ziggurat, the sanctum sanctorum of a Tower of Babel, a confusion of tongues and specialties, so far out on the everexpanding peripheries of knowledge that often one can talk to no one but oneself, in masturbatory idolization and isolation. After all, when one dotes on the gonads of South African grasshoppers, why show joy at the other bug-man's love for the bedbugs of urban ghettoes.

The university used to be the temple of learning and education. Thru sociologic forces and powers far beyond their own control, "research" has gone into the driver's seat; and "teaching" is hanging on to the underbelly. Classes have become larger. The eminent professor lectures to hundreds at a clip, and meets no one personally. Actual contact with the student is thru the graduate teaching fellows, themselves budding specialists with no interest in teaching. The universities need money; the only available money is for specialized research. When did money not pervert? The evidence is clear when one of our most famous teaching universities—at this writing—is boasting of the tallest building on its campus, even in the entire city, a building that does not have a single classroom. Why try to teach well, when a university president of another campus makes it an open declared policy that all new appointments, salary raises, tenures, are dependent upon published research; and when specifically asked about the superior teacher, liked by the students, but who does no research—the answer was a simple "He does not come in, if he is out; and he goes out, if he is in."

This change in emphasis at the University is having a striking and marked effect on the kind of Scientist. It is true that the relationship between teaching and research is a mumbled dogma that has not changed. I recall that as a college freshman in 1929 I was told about the wise balance between these two virtuous activities of the professor. But that person had published his doctoral thesis, and tho he praised his "research" in all of his classes, to my knowledge he essentially retired without ever publishing a second paper. Today, we hear the research professors tell each other so often that they really believe it— even the deans believe it—that to be a good teacher, you must first be a good specialist and researcher. Science students think differently. The humanistic idea of involving the whole man in the quest for knowledge, order and beauty thru the ennobling exposure to other men's accomplishments has been mostly replaced by the training of task-oriented technicians. The university professor is becoming ever more a specialist. As such, he is ever closer to that type we have called the idiot-savant. There are exceptions of course. I could name many. The problem is inherent in the very nature and structure of our universities, and can only get worse before it gets better. I refer to the "vertical" organization and administration of our campuses, a chain-of-command that is divided according to specialties, into departments that have developed high loyalties within themselves, and a high territorial imperative against any infringement of their autonomy by other departments. My interpretation of this situation will be borne out by the future historian who describes the problems of establishing institutes and centers of "environmental biology," or "ecosystem ecology" now starting on our campuses. These are "horizontal" integrative units that extend across departments, divisions and even colleges. I doubt if any of our universities will actually convert. New universities will have to arise. I have tremendous respect for the provincial profundity of the professor.

> *I wish . . . to repudiate C. P. Snow, who intimates . . . that scientists should be entrusted with the world because they are a little bit better than other people. My view, based on long and painful observation, is that professors are somewhat worse than other people, and that scientists are somewhat worse than other professors.*
>
> —ROBERT MAYNARD HUTCHINS et al.
> *Science, Scientists, and Politics. 1963*

Alan Harrington

LIFE IN THE CRYSTAL PALACE

London: Jonathan Cape. 223 pp. 1960

There is no book to my knowledge that describes the life of the individual scientist in the benevolent Social Welfare system that is evolving in government, industry and university. In lieu, I strongly recommend this book by Alan Harrington, who spent more than three years as a Public Relations official in the suburban headquarters of a great American corporation with many overseas affiliations—and gave it up. The life he describes, the life of the individual employee, bears so many similarities (not identities), not just to those of the scientist-technologists employed by industry, but to scientists in government and in the spawning grounds of these people in the research factories that are still called universities, that I am pained by the resemblances.

Alan Harrington writes simply, effectively, often straight from the shoulder, but objectively and without resentment and rancour. Only after finishing the book does one fully grasp its frightening implications. Here is a microcosm of trends in the social evolution of Homo sapiens. The System chooses, encourages, and thus breeds a type of individual who is excellently decent, docile, amiable, cooperative, compliant, and eminently mediocre. Of his own free will he picked out, called for, and has chosen a protecting blanketing benevolence, a kindly and paternal authority, a lifelong social and material welfare that frees him from all the struggle, competition, responsibility, anxiety and uncertainty that biologically evolved him in the first place. To pay for this he has, in his freedom, abdicated all right to be individual, different, responsible, aggressive, and sagely superior. Mankind is again at the parting of the ways. With the imminence of world peace, we will find which is the more immanent in the human soul: the yearning for security as an unimportant cog in a larger social whole; or a craving for quality that segregates the superior and has in turn guided the evolution not only of mammals but of all vertebrates in the course of the earth's history.

SCIENCE CONCEPTS

We must never conceal from ourselves that our concepts are creations of the human mind which we impose on the facts of nature, that they are derived from incomplete knowledge, and therefore will never exactly fit the facts, and will require constant revision as knowledge increases.

—A. G. TANSLEY

(in the Journal of Ecology 8:120. 1920)

The foundation stone upon which rests the entire structure of an orderly and rational science is the idea that we and the world about us are comprehensible to the mind of man. This by itself is an audacious assumption! As stated in the Preface, nature is not only more complex than we think. It is more complex than we c a n think. Nonetheless, men have gained considerable faith in their aspiring science by its reliability (as did Agamemnon, as he sacrificed his daughter Iphigenia, in Aulis, so that the winds would blow the Greek ships to Troy), and by the consistency and uniformity of its predictions. If they predict wrongly (as happens more often than the high priests care to admit), then hindsight can at once be rationalized, often with a resultant advance in science.

 "Reality is not what is; it is what the layman wishes it to be."

If a scientist is at all creative, it is not in the data that the technician-aide accumulates and manipulates, but in the concepts which he constructs. These concepts do not exist in the outside world, but only in his mind. What distinguishes the scientific approach from all others are these concepts. Indeed, natural science is constituted of conceptual systems by which the mind of man orders, connects, and explains the properties of natural things in space and time. A concept is nothing more than an idea, a mental creation, which makes comprehensible a certain group of facts.

The concepts give us a path t o w a r d s reality, without the promise of attaining such a rapturous goal. For you who are wolfishly impatient in your emotional pursuits, there are other concepts and other paths to a complete knowledge, that guarantee a speedy attainment of the craven goal. How many, for example, return to a fundamental faith in their later, declining years when time is obviously running out?

A few concepts are extremely important. Several are discussed now (listed below) as being fundamental to the adequate comprehension of science as presented in this volume. It is not to be thought that any one or several are exclusively and permanently valid. They each serve their purpose, rising and falling in favor with the changing fashions of the times.

1. *Mechanism*

2. *Determinism and Causality*

3. *Crowd Phenomena*

4. *Chance and Probability*

5. *Holism*

 ". . . science creates the real world in its own image."

It will be noted that REALITY is not considered as a concept worthy of discussion in this treatise! Reality is not what *is;* it is what the layman *wishes* it to be. The "facts" and the "data" of science are not reality; they are only the raw materials, from which concepts are created, which in turn arrange and rearrange those raw materials. Science may approximate an unknown reality, but it will never attain that reality. Science is a product of man, of his mind; and science creates the real world in its own image.

The mind is the great slayer of the real.

—Fritz Kunz
The Voice of the Silence

It will be noted also that TRUTH is not considered as a concept worthy of discussion. Science may approximate an unknown truth, but it will never attain that truth. Truths are as amiably nurtured by scientists today as they were by philosophers a century ago. When dealing with concepts, the difference between "scientists" and "philosophers" is but a semantic quibble. Thus it is peculiarly appropriate to quote Friedrich Nietzsche, from *Beyond Good and Evil, Prelude to a Philosophy of the Future,* Section 5, as translated by Walter Kaufman. Nietzsche writes "What provokes one to look at all philosophers half suspiciously, half mockingly, is not that one discovers again and again how innocent they are—how often and how easily they make mistakes and go astray; in short, their childishness and childlikeness—but that they are not honest enough in their work, although they all make a lot of virtuous noise when the problem of truthfulness is touched even remotely. They all pose as if they had discovered and reached their real opinions through the self-development of a cold, pure, divinely unconcerned dialectic (as opposed to the mystics of every rank, who are more honest and doltish—and talk of "inspiration"); while at bottom it is an assumption, a hunch, indeed a kind of "inspiration"— most often a desire of the heart that has been filtered and made abstract—that they defend with reasons they have sought after the fact. They are all advocates who resent that name, and for the most part even wily spokesmen for their prejudices which they baptize "truths" —and *very* far from having the courage of the conscience that admits this, precisely this, to itself; very far from having the good taste of the courage which also lets this be known, whether to warn an enemy or friend, or from exuberance, to mock itself."

1.   Mechanism

*I believe many will discover in themselves a longing for mechanical ex-
planation which has all the tenacity of original sin. The discovery of such
a desire need not occasion any particular alarm, because it is easy to see
how the demand for this sort of explanation has its origin in the enormous
preponderance of the mechanical in our physical experience. But nevertheless, just as the old monk struggled to subdue the flesh, so must the
physicist struggle to subdue this sometimes irresistible, but perfectly justifiable desire.*

—P. W. BRIDGMAN
The Logic of Modern Physics. 1927.

Perennially one of the most popular, valuable, reliable, and satisfying of the scientist's philosophies towards all life and the universe is
mechanism. Mechanism considers the universe and all that is in it
as a great machine, a physico-chemical system, explainable in terms
of matter and motion and their governing natural laws. The universe
is a very complicated but eventually comprehensible machine.

Adversaries of this philosophy take the stand that such aspects of
our life as the psychologic, esthetic, moral and religious will never be
reduceable to terms of the chemical and physical. They propose a
vitalism, which takes the view that there is some principle or element
in living things—call it mind, free will, soul, spirit, or entelechy—
which directs and intelligently manipulates the world of living things.

The history of mechanism, like so much else in science, dates back at
least to the Greeks and Romans. During medieval centuries, mechanism went in to total eclipse. Then came Sir Isaac Newton (1642-1727),
and the establishment of classical mechanics.

The appeal of mechanism to the biologist is readily understood. The
hard facts of physics and chemistry make a universe that the mind of
man, if not his hands, could look at, grasp, confine in his laboratory,
measure, and subject to tests and experiments. Prediction became possible. With prediction, the universe became orderly, a necessary assumption of scientific research. The same chemical elements found in the
inorganic Earth were not only found in the distant stars, but found to
compose all animals and plants. Here was a way out of the imaginary,
the artificial, the fictitious, the subjective, and the speculative. If the
biologic world was not directly reducible to a gigantic and confused
contraption of wheels and levers, pulleys and wires, at least some day
it would be reduced to analogues of these. Thru the first half of the
20th century, a majority of the men of science held to such a naive
materialism. It is now fashionable to refer to mechanism belittlingly,
to downgrade it in favor of more "sophisticated" concepts.

The crux of the issue perhaps lies in the separation of (1) mechanism as a complete interpretation of all life and the universe, from
(2) mechanism as one of several convenient scientific concepts, used
when found workable, but to be abandoned when need be.

2. Determinism and Causality

The idea of mechanism leads naturally into that of *determinism*. Since the machine-like universe is already in operation, in some way it must have started, been wound up, or otherwise initiated; and speculation upon this pleases those with an element of vitalism in their natures. (A supreme intelligence you see first started things). Furthermore, since the machine is already going, *the future is determined*. This is the faith of the determinist. There is no use fighting it, the die is cast. We should accept what comes. Such is a favorite philosophy of some deans and chancellors. They come in to office like an innocent inheriting a wound-up grandfather clock; and hope to all hell that it won't need rewinding or repair until they, not the clock, stop ticking. This deterministic attitude is much liked by scientists who have an element of the fatalist in them. ("Can't help it." "Why fight against the inevitable?") Determinism on the other hand is completely incompatible with "free will," a simple philosophy a scientist always holds for his own personal and intelligent behavior, but never for the other person who disagrees with him, even if his best friend. As a system which allows for acceptable prediction in the rest of nature, determinism is a most useful concept in the rest of nature, and will always play a role in the total scene.

Determinism leads naturally into *causality*. The intellectual essence of causality is a temporal relationship between two phenomena: an antecedent called the *cause*, and a subsequent called the *effect*. (I have yet to find the scientist who tells me that, when I ask about causality —but perhaps I travel in the wrong circles.) Knowing of one from sundry experiences, we can interpret the occurrence of the other. We can explain. We can predict. We can control. All-important activities these are in science, as soon as we establish a uniformity of sequence of phenomena.

The idea of causality I am sure dates far back in human history. When-you-eat-that-plant, you-get-pain-in-gut was undoubtedly a common cause-effect experience of our ancestors. Thereafter, the more such knowledge that was obtained, the more man could control his destiny. When such knowledge was concentrated in one person, he had the esteem of, and power over, the others. Thus, much of the respect for the contemporary scientist.

The idea of "cause" also has tremendous emotional significance to all laymen, and to the lay element in all scientists. The search for past causes seems to be one of man's chief psychologic delights. It helps to "explain," to rationalize the present. It has a soothing, cloying, satisfying effect on what would otherwise leave us uncertain, unhappy, in a quandary. How often do you hear an irritated "What causes . . .?", or a whining "Why . . . ?" Notice the relaxation and comfort that comes from "knowing." Such knowledge is a simple and terminal one-to-one relationship. The layman's curiosity however then gets laid. Those who "know," no longer ask why—or ask at all.

One of the most famous pitfalls of all logic lies imbedded in this compulsive urge of humanity, that known as "post hoc, ergo propter hoc" (*after* that, therefore *because of* that). *After* taking a placebo sugar pill we lose our headache, therefore *because of* the pill. So firmly engrained in the human mind is this way of thinking, so compulsive the search for single initial causes that I honestly believe that if this trait could be eliminated from the thinking of scientists, the entire field of scientific research would collapse over night. If you do not believe me, talk to scientists. Their interest in "causes" in natural phenomena is not one whit different, not one whit less dominant and compelling, than the man on the street—or even the woman of the street—who demands answers for questions about wages, prices, eating, and health.

The sophisticated scientist aware of this pitfall carefully refrains from too-hasty interpretations of cause-and-effect, and instead talks of "correlations," usually in high mathematical jargon. Sooner or later—sooner, when the positive correlations overwhelm him—he slips into causal interpretations. The literature in my own scientific field is loaded with premature acceptances of causality. Indeed, not too many years ago, ecology was actually defined as "the science of causes and effects." They really meant it. But we should not harshly criticize such men. As with all sin, it is easy to slip. It would be quite easy, for example, to stage a refined oceanographic experiment to prove that man can lower the sea level by dipping it out with a pail (provided one was not an astronomer, and accidentally ran his experiment when the lunar tide was receding.) The history of science is full of such quaint little episodes. What others are destined to be revealed in the future?

To a scientist, the sequence of cause and effect is, or should be, endless. An effect is but the cause for another effect. The cause was in turn caused by an earlier antecedent. An explanation is but the "pause that refreshes," so that he may next explain the explanation. Finding a cause really explains nothing. You go thru one door, only to find yourself in a room with a dozen other doors. This is the fascination of science. If you do want one Final Cause, I suggest that it is not science that offers it, but metaphysics and religion.

3.   Crowd Phenomena

Mechanism deals with material identities, that is, with individual people, with specific stars, with quantities of the chemical elements, and with mathematical descriptions of definite masses, or amounts of energy. All these phenomena are tied together into orderly deterministic cause-and-effect systems, that make a reasonable well-behaved universe for the scientist to study.

It was the physicist, just before the turn of the last century, that set loose a bull in the china shop of mechanism. By the time the bull finished his rampaging, the atom had been shattered, and Newtonian physics, once thought eternal, had passed into history. Quantum mechanics, the principle of indeterminacy, the calculus of probability, chance, and the so-called "statistical laws" dealing with large numbers of events, these have created a new look, even if they have not created a new universe. Mechanism is not dead; but mechanism took such rough treatment that the present concept is far less pompous and conceited, and now sits below the salt at the table of its betters.

 "Large-scale order is thus composed of small-scale disorder."

The physicist found that bombardment of atoms by subatomic particles scored direct hits in such proportion that probability alone could account for their average numbers, and chance determined which were hit. We can predict with great exactness the number of radioactive atoms that will disintegrate over a period of time. But who can say which atoms will crack up when? The significance of this view was that our old deterministic laws, with their direct cause-and-effect relationships, were found to pertain only to large-scale events, to *crowd phenomena,* to macro-situations. This regularity and statistical orderliness, when analyzed into its small-scale component parts, became disorderliness, random unpredictable irregularity. *Large-scale order is thus composed of small-scale disorder.* It was the unpredictability of small-scale events that threw scientists off their balance. In a sense they could not and would not tolerate it.

Science will N O T be involved with the behavior of capricious individuals within a "group." Such misbehavior is not a fit subject of study *for* science. There are various tricks for getting around this impasse, in addition simply to closing one's eyes to the "atypical" and the "abnormal." First, however, we should say that there was really nothing fundamentally new in this group-approach to phenomena around us.

One of the best known examples with inanimate objects is with games of chance, with tossed coins, cards, dice, horses, stock markets and all the playful games from Sodom and Gomorrah to Las Vegas. The most precise of scientific analyses cannot allow us to determine *which* coin will come up heads, and which tails, but we can predict very closely the *average* for large numbers. And thus the gambling casinos rarely, if ever, make the mistake of going broke.

In regard to men, no scientist will predict just which individuals will meet with tragic death. Yet experts predict very closely how many in a nation will die on the highways or in home accidents, during certain periods. Death and illness are referred to in the West as "acts of God," yet, by God, their insurance companies are so certain of the behavior of large numbers that they regularly pay dividends, pay beneficiaries, and uniformly make mountains of money for themselves.

 "How does the scientist experiment with one world?"

Astronomers, as they survey the multitudes of the stars in the Milky Way and consider the even more inspiring spiral nebulae, see the same order in disorder. It is the same macro-regularity arising from micro-irregularity that the physicist sees within the atom.

The problem that confronts the specialist in any one science is to judge which is an orderly *macro*-phenomenon, subject to "statistical laws," and which is a coincidental and disorderly *micro*-phenomenon, subject to "laws of the individual" much as a physician treats an individual patient—in the by-gone days of the family doctor.

The outdoor natural sciences, those operating at levels involving bio-communities of plants and animals, and complex vegetation types, have an especial difficulty. How does a scientist operate, for example, when he studies ponds, and finds that each and every pond is markedly different from any other? And yet he does not have enough "s a m- p l e s" (comparable to the millions of molecules of a gas) to treat statistically. He can, and does, study just *one* pond; and leans over backward to avoid any "generalization." This was customary in the first half of the 20th century. But in the second half of the century, when these sciences are trying dutifully to mold themselves in the image of chemistry and physics which alone is "science"—because this is where the shekels lie—there are other ways of studying lakes. You do not study their individuality, even tho it is as distinctive as the noses on some peoples' faces. Instead, you grind one all up, hamburger style. All hamburger looks alike. And then you analyze the "samples" now so-called, for fat, proteins, and an infinity of micro-chemical and micro-physical phenomena. I thoroly recommend this procedure, for one who is not too nosey about nature.

What happens when our "individuals" are not lakes, but islands larger and larger, an entire continent with all its resources plant and animal? And finally our one world itself. How does the scientist experiment with *one* world? This is where the contemporary philosophy of science falls flat on its face. Either the scientist or the layman must rise to this need, or historians of a future race will document our fall.

 "... 'exact scientists' ... are forever embarrassed by the non-conforming individualist."

Much as "exact scientists" (sic) are under the emotional compulsion to live only with the symmetry and order of groups and crowds, they are forever embarrassed by the non-conforming individualist. I do not mean the one that can be ignored, but the one who can *not* be ignored. The accident (you may be one) ; the un-caused; the demon *chance*. How does the scientist treat this disturber of the peace?

4. Chance and Probability

Large-scale order arising from small scale disorder, be it for atoms or for nebulae—a strange universe! What can be said for small-scale disorder? Does it merely indicate an inadequate state of human knowledge, which on further research will resolve itself into the patterns of mechanism? That is, does it represent the undiscovered areas of science? Is it a measure of our ignorance? Undoubtedly so, in some cases. Always so, says the orderly scientist, whose philosophic and emotional faith does not tolerate disorder. Or is there some fundamentally different opportunity for us? That is, can disorder be orderly considered? *Must* it be considered, if we are to understand nature? The man in the street talks of hazard, coincidence, unpredictability, probability, luck, fate, fortuitousness. He is forever "taking a chance," even to the magnificent minutiae of marriage, and its illegal alternatives. This idea of *tychism* (tyche, chance) conceives of chance as an objective reality. Some scientists are psychologically of such a nature that they demand the orderliness of mechanism, and say flatly that chance does not exist. Theirs is an easy way out of the difficulty. Other scientists have a different type of mind, and can work with phenomena of chance—in a way. These men have found that there are certain groups of characteristics which we will consider in the three paragraphs below, even tho the thoughts are not entirely separable.

A. Chance refers to a small unnoticed cause that results in a great important effect. For example, if we balance a cone on its point, and then release it, it will fall to one side. "Chance" decides to which side. Actually, if the cone were perfectly symmetrical, if its axis were perfectly vertical, and if it were subject to no other force but gravity, it would never fall at all; it would be balanced on its point. But there is an unmeasurable assymetry, a minute departure from the vertical, or the slightest air current;—and chance decides. If the experiment is repeated, the cone would fall in a different direction, for a still unknown reason. There are analogous situations in human history, where a chance saving of a life permits the reign of an Alexander or a Napoleon. Or consider the weather. Many still pray for rain who would never pray for an eclipse, tho once people did. The vagaries of the weather can result from an imperceptible difference in temperature at one moment in one place, which sets in motion an extraordinary train of events, a whole tornado even ravishing the coastline of an entire continent.

B. Chance refers to a phenomenon resulting from a great complexity of unmeasurable and constantly changing causes. The pattern of rain drops, the distribution of molecules in a gas, the mixing of two liquids, the distribution of playing cards in a pack that is shuffled, all these phenomena are unpredictable in detail. Think of all the energy of the human race that goes into bridge playing at the card table, with a constant and not necessarily large amount of sane intelligence pitted against madly fluctuating elements of chance, and results that are ascribed wholly to human intelligence, or its absence thereof. All such events cannot be repeated exactly, even tho the constituent elements appear to be the same. We say blithely that "chance" is involved.

C. Chance refers to a phenomenon that occurs when two systems, normally foreign to each other, react with a little incident that then has great effect on each. For example, an important statesman of one race is in the habit of walking each day to work along a certain street. A laborer of another race is a window washer, who quite accidentally dropped his bucket one day, which landed squarely on the pate of the statesman, and killed him. The laborer was killed by members of the statesman's race; a race riot resulted; their respective nations declared war; the world was involved in an atomic conflict. All because pail hit pate by happenstance.

The situations described above are no less "chance" than the successful meeting of egg and sperm which carried the coincidental cargoes of genes that gave rise to the Buddha. You yourself are merely a minor fortuitous event.

This concept of chance as a reality in the natural world lends itself to description by a system that has come to be known as the "calculus of probability." In its simplest form, probability is expressed as a fraction. For example, the probability of drawing an ace from a deck of cards is four chances out of 52, or $4/52$, or $1/13$. Probability is considered the relative frequency with which an event does occur within a certain class of events. It is interesting to note that the actual occurrences are mathematical facts, but the definition of the class of events involves a s u b j e c t i v e decision. And after the probability *is* stated (simply a mathematical description: comprehension and understanding of the phenomenon are not involved), the calculators of the calculus get out: scientifically far too complicated for further scientific study. Science cannot eliminate the philosophic!

> *Lest man suspect your tale untrue,*
> *Keep probability in view.*
> —John Gay (1685-1732)

5. Holism

These small-scale phenomena (analogous to one particle of gas, or one minor planet of a minor star of a minor spiral nebula) are not necessarily as simple and easy to recognize as *a* molecule, *an* individual organism, *a s*pecies, or *a* planet—except in that sole instance where I or we are the phenomenon. But how can we be, and be part of, and be divisible into, so many different entities in nature? . . . Are these entities equally existent and studiable in nature? What do they have in common?

One of the grandest and most creative of concepts, especially in the fields of biology, is *holism*. Holism allows us to unite, and to draw enlightening analogies among, a wide variety of phenomena that would otherwise be diverse and unrelated. Holism is the doctrine that the studiable units in nature are the more-or-less integrated wholes, rather than their constituent parts unrelated to each other.

The idea has a strange historical development. It was early accepted in physics and chemistry, where "systems" and "steady states" have long been studied. It is related to the "regional concept" of the geographers. In human history, Toynbee describes "societies" as the intelligible field of study, of which he recognizes a score in the 5-millenia history of civilization, of cultures that have risen and have fallen, or that will fall. In philosophy, ethics and religion, the idea of "emergent evolution" sometimes goes into realms that confuse the scientist. In biology, zoologists and botanists have followed very different paths. Zoologists have had little difficulty in comprehending the various "levels of integration." The very nature of their material makes the concept part of their frame of thinking. For example, poorly integrated colonial protozoans, coelenterates and sponges on the one hand, and highly integrated societies of insects on the other hand, have made commonplace the ways of holistic thinking.

Botanists, however, have seemed particularly immune to the idea. This situation is understandable when we realize that 95 per cent of all botanical science revolves around (1) individual plants, as distinct and separate as individual rats, skunks and human beings, as pets and friends; and (2) species which at least until the mid 20th century were strict segregationists, proud of the moral purity of their race, rigidly maintained since they left not the Mayflower but the Ark. Students of vegetation and of plant-communities have been recruited from the ranks of these taxonomic botanists. Utterly confused by the the merging nature of Vegetation phenomena (and when not foolishly arguing as to whether all Vegetation is composed of either intergrading continua, or completely distinct communities with real or imaginary stone walls between them), the Vegetation Scientists have talked about orgs, super-organisms, epi-organisms, social organisms, complex organisms, and quasi-organisms, with these terms quoted and misquoted in elegant nomenclatorial superfluity. The chief hindrance to grasping the idea of holism is our stubborn determination to project upon the natural world that particulate nature that each of us possesses. So let us now imagine ourselves merely as one cell of a colonial protozoan, as one bee, or as an algal cell of a lichen, and see what happens as we analyze the concept in its basic essentials.

The following 12 thoughts are worthy of separation. 1. Elements of lower categories, as chemical elements, and individual bees, unite to form patterns of higher complexities, as molecules and bee colonies. 2. The new whole is something more than a mere sum of the parts, even as H_2O is something more than the sum of two parts of hydrogen and one of oxygen, even as you are something more than the sum of the chemical constituents in your body, worth 98 cents when I was a student, but inflated to the price of four dollars or more these days. 3. The characteristics of the parts may or may not be present in the whole. The characteristics of hot air may or may not be typical of a man, such as found in the lungs. 4. New phenomena or activities emerge, as characteristics of the whole. These phenomena do not reside in the elements, but in the whole formed by them. Watch a race riot, in Calcutta or Detroit, and note the mob spirit that emerges, characteristic of the group, and within which the individual is but an unthinking component. 5. The organized whole, hereafter referred to as an "Organism" (something organized), is a comprehensible entity in nature, and may be studied as a unit without actual recognition of its parts. The ordinary driver of an automobile should not be disturbed by one loose screw in the back of his vehicle (unless indeed it is a tychic trigger factor, like the window washer's bucket that would completely destroy the equilibrium of the smoothly functioning holistic unit). 6. The Organism is adequately self-sufficient, stable, and generally tends to perpetuate itself, essentially unchanged, in nature. This is true of the "conglomerate corporation" of this economic era, even tho the individual human beings are manipulated like the merest puppets, as both producers and consumers, for this social Frankenstein. 7. Organisms exhibit various degrees of integration, some being well-developed as a bee colony, much more so in a single human being. Others are very loosely formed, as scattered weeds in an old field. 8. Organisms may unite to form larger Organisms of a higher category, as individual organisms group themselves to form a social Organism, such as in a small successful family business. 9. Organisms are of varying complexity and may be considered as parts of an ascending sequence of categories, e.g., atoms, molecules, individuals, and bee colonies. 10. Organisms often exist in an organized relation with other Organisms, e.g. the clover-bumblebee-mouse-cat-spinster complex of Darwin. As he explained it, spinsters kept the cats, which fed on the mice, which ate the bumblebees, which pollinated the clover, which was so necessary for British agriculture. One might add that this was the clover, which produced the forage, that was fed to the steers, that produced the beef, that nourished the navy, that created the British Empire. With our changing moral standards, this complex linear system has been short-circuited. With direct contacts between the Navy and the spinsters, the old time spinster has vanished from the British scene. Fewer cats; fewer mice; fewer bees; and so on d o w n, even to the Navy and the Empire. If there is a moral to such tale: the spring of pursuit ends in the fall by attainment, of empires as well as men. Post hoc ergo propter hoc? 11. One element may play a role in more than one Organism, as a human being may function in several distinct social groups, racial, religious, economic, civil, familial, sexual. 12. The universe itself is a very complex system of interrelated wholes of many ranks and levels. According to their degree of organization and their distinctness in nature, so is it desirable to recognize holistic Organisms for scientific investigation, and as legitimate subject matter for the natural sciences.

We have in the preceding pages discussed various concepts, various intellectual ideas, that exist in the human mind and that facilitate us

 "Concepts are games we play with our heads; methods are games we play with our hands . . ."

in developing that body of knowledge which today we call Science. We now move on to other pastures which we group under the designation of "methodology," of procedures which can be routinized for the benefit of technicians. It is not to be thought that the dividing line is sharp between concepts and methods. They are merging and overlapping phenomena. Concepts are games we play with our heads; methods are games we play with our hands, which at times are so handy they can be played without a head.

METHODOLOGY OF SCIENCE

*A technique is merely a recipe for use in a constantly recurring situation
that has lost all its problematical aspects. Most techniques are only habits
with which to avoid thinking. You cannot solve a new problem with an
old technique, for if you can, it is not a new problem . . . The intellectual
. . . tension that is . . . (a great man's) answer . . . (to a new problem) is
not only an essential ingredient to a great work, . . . it is probably the
greatest of its ingredients.*
—WILLIAM W. IVINS, JR., Emeritus Curator of Prints,
Metropolitan Museum of Art.
(From an article on prints)

Methodology is the field of knowledge concerned with the orderly
description and evaluation of techniques, procedures and methods.
Each science has its own methodology, even tho sometimes the same
methods are used in several sciences. Furthermore, there are special
methods that pertain to subdivisions of any one science. These are detailed specific techniques beyond the scope of our present interest,
as are also those which made Rube Goldberg famous.

At this time, we shall consider methodology only in a general sense,
as it pertains to the entire subject of science. A science is to a great
extent the product of the methods applied to it. We find only what we
look for. We strain out only what our strainer is designed to detain.
We net only those fish neither too small to escape, nor too large but
that they break the net. We photograph only what passes thru the
lens and affects the film. We measure only what is measurable. Indeed, science is but an artifact of its methodology. Nothing more.

For these reasons, we must never forget that methodology is concerned with the known. What is unknown or unguessed, remains unknown! We never know the unstrained, the unnetted, the unphoto-
graphed, the unmeasured.

A method is a systematized procedure and a logical order of activities for the accumulation of knowledge. As such it can be one of the
finest products of the human mind. It is a "tool" in a figurative sense,
an intellectual tool, that permits the orderly amassing of data with a
remarkable efficiency of time and energy. A method should be treated
with respect, for it is powerful.

Methods are designed by intelligent people (who use them intelligently). They can be used by unintelligent people (who use them
unintelligently). In this sense, no method is fool-proof, that is, proof
against being used foolishly by fools. The more complicated the procedures, the more foolishly they can be used by intelligence levels below themselves. In this sense, methodologies are strewn with pitfalls
for every one, and the scientific literature is strewn with the wrecks

of falls into them. Unfortunately, the wrecks often present such strange bizarre and futuristic appearances, especially when draped in mathematicese, that they are looked upon with respect and awe by the uninitiated. I am not at all sure that I myself can recognize wreck from rectitude in fields not my own. I don't expect a layman to be able to do so. Herein lie the dangers of the age.

Methods are a necessary adjunct to scientific activity, in the same way that habits are necessary to our daily living. Their application always deserves the finest and best thinking of which we are capable. In the following several pages we will consider different facets of the general problem. These facets are not fractional parts of the subject. Rather, they are "views," like taking pictures of a crowd from different spots and different times and thru different lenses. The "lenses" we will use are the following:

1. *Methodology and Concepts*

2. *Methodology and Nomenclature*

3. *Methodology and Logic*

4. *Methodology: The Useless and the Useful*

5. *Methodology and Psychology*

6. *Methodology and Instrumentation*

7. *Methodology: Observation versus Experimentation*

8. *Methodology and Sampling*

9. *Methodology: Averaging versus Ordination*

10. *Methodology: Biometry and Variation*

11. *Methodology: Mathematical Formulae; Formulation of Laws*

12. *Methodology: Systems Analysis*

 "I would sooner trust a good mind without a method, than a good method without a mind."

For all their value, the application of a method, alone, is not science, any more than a pile of bricks is architecture. I would sooner trust a good mind without a method, than a good method without a mind. The genius of inventiveness, so fecund in the scientific mind, may be sterilized by a poor method, while a good method may serve as fertile soil for its further growth and development.

W. I. B. Beveridge

THE ART OF SCIENTIFIC INVESTIGATION

New York: W. W. Norton. 171 pp. No date (post 1948)

Professor Beveridge, whose specialty is animal pathology, has written a book that is the distillation of many years not only of his own research and of the training of graduate students, but also of the deeply intellectual interest in the lives and works of great scientists of the past. It abounds in well-chosen quotations that are a gold mine of their own.

I most warmly recommend this book, both to the layman who seeks to understand the events that happen at the frontiers of knowledge, and to the young graduate student who thinks of embarking on a career of scientific research. The book will, of course, be meaningless to those hordes of students that are acquiring graduate degrees in science only because society demands the degree for their future routinized jobs, and who have neither the intellect nor the interest to indulge in the intellectual inquisitiveness of the independent researcher.

One should note the title: Scientific Investigation is an *Art*. Research itself is not a science; it is an art or craft. As such, Professor Beveridge is strongly man-oriented. All the plates in the book are portraits of well-known scientists, 16 of them. Furthermore, the headings of the eleven chapters are themselves indicative of his respect for qualities and characteristics of *men*, and not just of parameters that can be turned into numbers and fed into a digital computer. The chapter headings are: Preparation, Experimentation, Chance, Hypothesis, Imagination, Intuition, Reason, Observation, Difficulties, Strategy, and Scientists.

The book closes with a bibliography of 109 titles, itself a rich source of materials oriented toward outstanding scientists, their own ideas, and what they themselves have accomplished.

1.   Methodology and Concepts

Concepts have already been discussed (pages 21 to 34), as being the very stuff of which science is constructed.

At this moment, we must think of concepts again, as themselves methodologic tools. A concept *is an intellectual method,* supplying frameworks, routines and procedures with which to organize and handle the perceptions that enter the mind. These ideas, these intellectual constructs, can be immensely fruitful and important devices for developing a science.

In history, conceptual methods—in this sense—antedated all others. This situation is especially apparent when we review the entire realm of Greek knowledge. Their thinkers and philosophers were involved with ideas primarily, and at times they did not allow objective experiences to conflict with or to interfere with conclusions otherwise derived. This situation was carried forward and developed into absurdity through more than a thousand years in western Europe, when a dogmatic Aristotelianism was upheld and defended, even when factual experience of the outside world was in open conflict with it. The burst of Renaissance thinking eventually killed such dogmatism, but only after several decades of rear-guard opposition—from the Church.

In the 20th century, the pendulum is swinging in the other direction! The material, objective and external characteristics of the methodologies are believed to be not only all-important but exclusive, s o l e , attributes of the methods. The view is sometimes referred to as a "naive Victorian realism," an attainment to an external and objective truth and reality that is quite independent of the emotional and subjective components of human nature. If you must use the word reality, then "reality" *is the method,* the mindless method.

A very serious problem inherent in every method is what might be called the Obfuscation by the Observer. In simplest terms, this hypothesis proposes that any attempt to investigate a subject matter profoundly alters the material which is being investigated. The bacteriologist who must stain his bacteria to see them, will never know whether they blushed as he observed them. We cannot wade into a pool to study its mill-pond surface, even as we can no longer study the atom, after we have destroyed its nucleus to find out what was in it.

In summary of these thoughts, we take the stand that every important scientific method, whether it involves our sense perceptions, or certain instruments for recording perceptions, or certain changes and controls over the conditions, or in the mathematical manipulation of the data, every such method on analysis will be found to i n c l u d e components that are purely ideas and concepts. These human components are fundamental limitations to the method, and must never be lost sight of when the method becomes an ingrained fixed and comfortable habit of thought.

2. Methodology and Nomenclature

By nomenclature, we mean a system of naming as applied in any science, especially in connection with classification. Nomenclature, purely and only as a form of *naming,* is the simplest and most basic of methodologies. It is often—too often—the only methodology.

Insofar as science is a search for knowledge, and the systematizing of this knowledge, we seek understanding and explanation of phenomena. When a child asks "What is it?", and we give it a name, he is satisfied. So also, in science, the very same trait can be discerned. We note two variations of this very human tendency.

The first variation is to give the unknown phenomenon a long and mystifying name, derived by tradition from Greek or Latin roots. There is a strange and amazing satisfaction in naming the unknown. It no longer seems unknown, even tho *we* know absolutely nothing else about it. Note how often in the scientific literature a phenomenon is named, even if no significant description follows. How often, in a social gathering, one is introduced to another, or one asks "What is your name?" Name known; satisfied; one launches into a conversation about the weather, or oneself. Never a further thought or question about the person we have just met.

The second variation is to give the unknown the name of something known, together with a qualifying adjective. By this device, the unknown object borrows some of the attributes of the known, together with an adjective which seems to "explain" all its unknown properties. Thus, an avocado pear, when first named, was something like a pear, and something not like a pear. For many people to this day, an avocado pear is not much more.

The naming of a phenomenon is an essential first step in scientific research, even as it is in the development of language itself. For the scientific nomenclaturists, it i s science.

Peter Achinstein

CONCEPTS OF SCIENCE. A PHILOSOPHICAL ANALYSIS

Baltimore: Johns Hopkins Press. 266 pp. 1968

This volume is not, as the title might indicate, a survey of the field of Philosophy of Science commensurate with the scope of this book of mine. To the contrary, it covers but a bare handful of the topics in the Table of Contents on p. xv. Nonetheless, Achinstein's contribution—by a philosopher, for philosophers—can be used as a valuable and interesting perspective upon Philosophy of Science contemporary with my own offering. It represents the combined wisdom of a professional group, replete with the tensions, conflicts and disagreements of a self-sufficient, academic system, acting in isolation of and without impact upon other academic groups, most certainly upon the world of Science itself.

The book is well organized, and the topics logically developed. Chapters one thru 3 involve *terms and definitions*. In this section, the author promotes his own position which stresses a key concept of "semantic relevance" (altho nowhere is there a clearcut definition of the concept, while one wonders why semantic relevance should not be ever and always essential in all philosophic positions). He finds inadequate the doctrines of other philosophers influenced by the history of science and by the empiricism of Logical Positivism which finds "truth and reality" in sense data and experience, free of all value judgments. Chapters 4 thru 6 involve theories, with their dependence on terms. Chapters 7 and 8 involve *models* (including analogies), with their dependence on theories.

For two thousand years philosophy, as the queen of knowledge, mothered and led other fields even when they became relatively independent of philosophy herself. Now we are witnessing the strange spectacle of a complete reversal of peckorders. "Science" has attained dominance, and "philosophy" is too ready to admit that it is this new upstart which has attained to truth and reality. Acting now as the menial, philosophy is content simply to watch, observe, describe and interpret, not nature, but science. I am sure there are some who, upon reading this book, get the uncomfortable feeling that the philosophers, craving to know about this road to reality, have looked up definitions in dictionaries. Thereupon they are entranced with the wonderful world of words, lost to analytic lexicography, devoted to terms and their meanings, comparing contrasting classifying them in endless permutations and combinations. Yes, a few individuals rise above this technician-level. So it is with many fields.

3. Methodology and Logic

Logic is rationalized desire.

—reputed to a Greek philosopher

Logic is the field of knowledge concerned with sound and exact reasoning or thinking. As such, it is pursued and wooed by the scientist with all the passion and ardor of Tannhäuser in the enchanted caverns of the Venusberg. When not wooed, but simply acquired by fiat—which is now the more often—it is displayed as the gaudiest jewel of the crown he claims by Divine Right. Never question a scientist on his right to reason any more than you would have questioned Louis XIV on his right to rule.

It is a wise man who is not roped by his own rationalizations. Said Benjamin Franklin (1706-1790) in his *Autobiography,* when he finally left his vegetable diet, and returned to eating the meat and fish that he so enjoyed, "So convenient a thing it is to be *a reasonable creature,* since it enables one to find or make a reason for everything one has a mind to do."

 "Never question a scientist on his right to reason . . ."

It is conventional to consider logic as that branch of philosophy concerned with the "true," in the same manner as ethics is involved with the good, and aesthetics with the beautiful. Further than that, we had best not advance at this time, for logicians are by no means agreed upon the scope, content and emphases of their field. Any further discussion would involve us in terminologies that would vie even with those created by the ecologists. These are "value judgments," a term which upsets the scientist.

"*The* scientific method"—all too often it is a substitute for imaginative thought—is a phrase sometimes used to refer collectively to a rational series of processes and steps by which scientific "laws" are supposed ultimately to be established. It is thus a specific and favored application of logic, considered by the scientist to be central, as a central heating system would be, for all the hot-air conduits to all the rooms of the sprawling mansion of science.

1. The first step in The Scientific Method is the *recognition of a problem*. Unless we realize there is something we do not know, unless we are curious beyond the point of simply asking "why?", unless there is an obstacle or a difficulty, there is no development of science. It is amazing, and discouraging, how frequently this first step is either side-stepped, or if faced is not answered. Graduate students, when first confronted with the need to plan research, are frequently stumped. In all their training, they have never been taught to ask questions, but

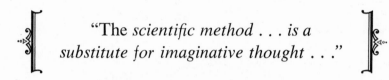

"The scientific method . . . is a substitute for imaginative thought . . ."

only to give answers. They have been trained to believe that science h a s the answers; it is impolite for t h e m to ask questions. And then they soon learn, from the research professors, not how to ask questions, but how to evade asking them, and still conform to the rituals of the religion. The methodologies of the science, in their specific applications, allow all too easily for the simple accumulation and treatment of "data," with very little more thinking than needed by an assembly-line worker in an industrial parts factory. The recognition of a problem will always remain as a superior intellectual expression of the human brain.

2. The second step of The Scientific Method involves accurate observations and measurements of the things involved in our problem. This is the time when we accumulate raw data. Often mountains and mountains of it. Once the trick is learned, it is like a drug habit, craving ever more and more, yet quite unable to break the hypnotic effect. I have known many, many scientists, both in government and academic circles, with such an overburden of data in their files that neither they nor their administrative superiors have the scientific creativity to use it. Such is the common way of the common man.

3. In the third step, the data are organized and analyzed. Similarities and dissimilarities are noted. The data are arranged in some meaningful order and classified. From the classification, generalizations can

be made as to common properties of different groups of data. There is simply no end to the games one can play with collections of data, even as one can with collections of buttons, safety pins, or postage stamps. I have been asked to evaluate many Project Proposals where such games seemed to be the end and goal of the entire activity. I am not disparaging such activity. Nuggets of gold are frequently found amongst unprepossessing gravels. I would simply like the actual gravels evaluated, as well as the possibility of gold.

4. Now an hypothesis or hypotheses are formulated. These are tentative solutions or interpretations of the phenomena, that seem reasonable in the light of the information available. The more unimaginative the mind of the scientist, the less creative he is, the more of a data-gathering hack he is, the more shy he will be of coming out with any firm neck-protruding hypotheses. Such reticence can easily be rationalized. He may claim that his data are "insufficient" (and indeed they often are—being simply busy-work to fulfill the requirements of a research grant). He may claim that the data are "conflicting" and support divergent hypotheses. He may claim that other unknown and unmeasured factors were operative. There are no ends to the graceful side-steps in this mincing minuet.

Steps 2, 3, and 4 are often collectively known as *induction*, that is, reasoning from the particular to the general. We associate this mode of thinking with Sir Francis Bacon in the 17th century, who stressed it to the essential exclusion of other thinking.

5. Assuming the proposed hypothesis is valid, the scientist then considers what other situations must follow. That is, he reasons out the consequences of this tentative solution. "If so and so is true, then such and such must also occur and be true." This is reasoning from the general to the particular, and is known as *deduction*. Deduction was the type of thinking used almost exclusively by the Greeks. They would sometimes conceive a theory, and reason from it, never checking with observations around them. Indeed, it was more than a thousand years before people had the courage, or the curiosity, to dispute some of these deductions which, in the meantime, had crystallized as "truth." Imprisonment or death was often the penalty for questioning such established truth. Even today, hypotheses can become theories, and theories can become "truth"; (e.g., animals are not intelligent—only because the original hypothesis requires it) and when some brash young scientist dares to observe facts contrary to "reality" he may not be burned at the stake (if he gets into print at all), but the oldsters try to approximate the act in published effigy.

6. The final step in this procedure of The Scientific Method is the verification of the deductions. That is, "If such and such should occur, d o e s it occur?" If it does occur, we get bolder and frame a *theory.*" If the theory becomes firmly established, we frame a "law." The word law is most unfortunate for the purpose, for it reeks with anthropomorphic connotations. Scientific "laws" are only man-made generalities that are by no means binding on nature, but "hold" only for the situations under which they were formulated. They are neither sacred, nor "true."

 "Genius knows no method . . ."

It is not quite honest to imply that the best of our scientists have reasoned clearly and accurately and progressively in these six stages of The Scientific Method. More likely than not their observations and their thinking have see-sawed back and forth between induction and deduction. Furthermore, there has undoubtedly been a very heavy component of, and jolts from, serendipity, exhaustively explored intuitions, persistently pursued hunches and fortunate conclusions. Genius knows no method, tho it leaves a spoor-method for drones to follow.

> *The mind of man most loves those errors and delusions into which it has become self-persuaded, and is most fanatic concerning the irrationalities and supernaturalities to which it has bowed its own reason.*
>
> —Sir Richard Burton (1821-1890)
> *The City of the Saints*

4. Methodology: The Useless and the Useful

The last word in ignorance is the man who says of an animal or plant "What good is it."

—reputed to ALDO LEOPOLD

There is another progression in the methodologic development of science which is certainly not logical (in one sense), but is factual (in the sense that it does actually occur). It is important for the layman to understand this progression, for it is generally the scene of bitter and acrimonious controversy between scientists of the early stages (who never move on to the later stages), and scientists of the later stages (who never had passed thru the early stages). There are personality differences between the two groups, and their inability to understand each other is the source of some of the more pathetic internecine squabbles of academe.

1. The first stage in the development of a science is *understanding*. At this time we comprehend, or think we comprehend, the nature of the phenomena, in terms of the *laws* just discussed.

2. An understanding of these laws allows *prediction*. Prediction, or the forecasting of future events, is one of man's fondest desires and pastimes. In history, mankind has predicted long before he understood. The history of astrology, for example, is one of the finest examples of this compulsive trait in human nature. The arts of divination are legion, and the literature on them is as delightfully imaginative as any collection of fairy tales. True, man often predicted wrongly. Heads were lopped by the hundreds for such mistakes, but tongues in other heads were always ready to keep on predicting. In the mid-20th century, we are far better equipped to predict. (Think of the prognostications of the Weather Bureaus.) In the 1960's, it was a favorite game to make predictions for the year 2000, a number which seems to have hypnotic fascination for the diviners. The best of our computers were called in to the act, so as to make the prophecies "scientific," and technologically "infallible" (in a neo-Papist sense). As any child can realize, however, all such predictions are merely extensions and prolongations of short-term recent trends. That they should prolong themselves uniformly and consistently is the absurdity. They never have, in all past history. Furthermore, when these prophecies show up extremely undesirable situations, such as an infinitely increasing human population, it is a reflection on the non-intelligence of humanity that our thoughts are directed merely to f e e d i n g that increase (clearly impossible in a finite world, even if food alone could create a civilization), rather than p r e v e n t i n g that increase.

Understanding and prediction together constitute what is often referred to as "basic science," "fundamental science," "useless science" (except for predictions arising from boards of economic advisers), or "pure science." The term pure is unfortunate, for it implies that other aspects are impure, or at least contaminated by usefulness. Yet that opinion is exactly the attitude of many who call themselves pure scientists. Herein lie the "natural sciences," as in the Life (magazine) Nature Library. (See essay below.)

3. A grasp of understanding and prediction often allows man to *influence* the phenomena of science. By influence, we here refer to our ability to create and modify some minor phenomenon which in turn changes, redirects or alters the situation or the course of events. In the social sciences, more and more is being heard about PLANNING, especially since the unplanned megalopolises are incontrovertibly mistakes. It is interesting that the very word implies something on paper, rather than something implemented in fact. In the biological sciences, it is convenient to consider these activities as constituting the realm of the MANAGEMENT sciences. There are well defined fields of vegetation management, forest management, range management, wildlife management, pasture management, watershed management. It is as tho we lay our finger on a scales relatively in balance, exert certain relatively slight pressures, and throw the scales into a new configuration.

4. The final stage in this logical sequence from understanding, thru prediction, on to influencing, is when we step in and completely *control* the phenomena involved. In such cases we have totally artificial, not semi-natural conditions. Agriculture belongs in this category; and it is traditional to refer to these fields as the CULTURES. Apiculture, aviculture, silviculture, arboriculture involve phenomena in which man exerts relatively completely control. How logical some of these controls will prove from the standpoint of the future of mankind however, grieves many of our greatest thinkers. Perhaps we shall never know, for if we do not first make our Total Environment a completely unfit habitat in which to survive, we may work out the techniques for a very artificial anthropiculture. The results will be of interest to the paleontologists of some future species of the planet—if the planet survives.

Steps 3 and 4 collectively constitute the "applied sciences," the useful arts, engineering and technology. Herein lies "science" as in the Life (magazine) Science Library. (See essay below.)

Actually the distinctions between the basic and the applied are not only artificial, but misleading. Altho superficial andor technologic advances in the practical fields can often be accomplished by keeping within the last two steps, true advances have always depended upon the findings and discoveries from free and unfettered basic research. In turn, the freedom of basic research is not hindered by dealing with phenomena or materials that may also be of practical value. The distinction is actually a state of mind, depending upon a personality type. Each side has a big bag of rationalizations to "prove" the value of its own existence. The basic scientist is a lover of knowledge, oriented towards long-term and even practical developments in space and time. The applied scientist is a job-holding short-term realist, oriented to material comforts and the Gross National Product. So sharply are the lines drawn between the two clans (for example, we have two professional societies of entomologists, one the economic bug-men, the other the "pure" entomologists, with fundamental differences in their sociological behavior patterns particularly at their annual meetings) that anyone who by his own nature can bridge the gap, is suspect by both sides. The ultimate insult of the applied sci-

 "The overwhelming hypertrophy of technology—one of the most subtle and disastrous intellectual collapses . . ."

entist to the pure scientist who would invade his side (and threaten his own apple cart) is to be labeled a "theoretical professor." The pure professor who sees a colleague consulting for industry has a sigh for the one lost to prostitution. An applied scientist going to academe is simply a lost nut in a nunnery. Perhaps these distinctions were more apparent before 1950. The overwhelming hypertrophy of technology since then, i t s adoption of the name "science," and the atrophy of basic science—a true withering on the vine of the fruits of the future —may yet prove to be one of the most subtle and disastrous intellectual collapses in the history of the human race.

LIFE SCIENCE LIBRARY

Life Science Library (The word life refers to *Life Magazine*) comprises the following 26 titles: The Body, The Cell, Drugs, Energy, The Engineer, Flight, Food and Nutrition, Giant Molecules, Growth, Health and Disease, Light and Vision, Machines, Man and Space, Mathematics, Matter, The Mind, The Physician, Planets, The Scientist, Ships, Sound and Hearing, Time, Water, Weather, Wheels, and A Guide to Science and Index.

New York: Time Inc. 200 pages each. 1963-1967

This series, translated into ten languages, is unquestionably the most prestigious and influential single force in publishing history for the popularization of its subject matter. I would disagree vehemently with its title. This is no "Science Library," but a "Library of Technology, including its foundations in science." Technology rules the roost, and crows about its crown. The mis-titling not only reflects the confusion in the public mind between science and technology but, because of the influence of this series, will probably fertilize and establish that confusion in the English language. The stress on technology is intentional, and there is no criticism except on grounds of semantics. In the final Guide and Index, it is stated clearly in the opening lines, that the series is an "introduction to the world of science and its applications in human society." This is what science *is*—to everyone except the true man of science.

There are three Consulting Editors to the series, whose names are prominently displayed on every title page, men of extraordinary stature well deserving of this recognition, and authors in their own right: René Dubos, microbiologist and pathologist; Henry Margenau, nuclear physicist and philosopher of science; and C. P. Snow, scientist and novelist, famed for spotlighting the gulf between the "two cultures" of science, and the rest of us—a gulf that this series of books should markedly lessen. I hope these gentlemen will leave with their Wills and in their private papers a candid discussion of their actual roles in the preparation of this series, the extent to which they were consulted, the degree to which their asked-for advice was taken, and the extent to which their unasked-for advice was not taken.

The plan and format of each of the 25 volumes is a masterpiece of the art of printed communication, and leaves far behind the dreary run of poorly styled and designed books on science in which our students usually are forced to dip. It is accepted that the Life series has almost unlimited access to funds and to scientific, literary, artistic and photographic talent, but even without such resources, textbook writers and publishers can benefit by a perusal of this series. Each volume is composed of eight textual chapters, each with a following supplementary independent yet related "picture essay." The textual chapters are by the major author or authors, generally a team of a scientist and a journalist (even for those scientists who c a n write).

The margins are broad, and filled with pertinent diagrams and their legends. The picture essays, by the "Editors of *Life*," are predominantly photographs, largely in color and with a minimum of text. Many of the volumes start with historical discussions that will become permanent contributions to the strange and fanciful roots of science in lore and legend.

The final Guide to Science and Index follows a different plan. It has eight major chapters dealing respectively with physics, chemistry, microbiology, anatomy, mind and body, technology, geology, and astronomy. Then there is a large fold-out chart, with 41 topics listed on the ordinate axis in groups of five each (was a 41st added as as an afterthought, and if so, which one?) and the 25 volumes on the abscissa—a most convenient key for finding which subjects are discussed in which books. A 92-page index completes this volume, and supplements the separate indices that are in each volume.

Notwithstanding, since these volumes are purposely designed to sing paeans for "science and its applications in human society," for our freedom from disease (tho increasing millions die of disease), for our ample food (tho increasing millions die of starvation), for the way we have conquered the environment and twisted nature to do our bidding as the lords of creation, perhaps it is a bit unfair to judge this series from other than such man-arrogant and technology-worshipping stands. It may be said there are two "sciences": one is the world of technology, the glittering panoply of a temporary and self-acclaimed mastery on a limited spaceship, a cocky ecological high-jacker; the other is the world of a Total Ecology (far far removed from the simple little courses that were taught under that name in the first part of the century). This latter is the world that is scientifically concerned about the animal nature of man himself, burgeoning human populations, overuse of natural resources, pollution of the environment, the quality of human inheritance and of the environment, and the attainment of a reasonable equilibrium within the unity of man-and-his-total-environment. When these 26 volumes are measured against such ecologic standards, they are found to be shockingly deficient. True, one can glean a reasonable assemblage of sound ecologic statements—the proverbial needles in the haystack. The word ecology itself (now commonly used in the news media, and even in financial journals) is unbelievingly indexed as being discussed in but one volume! (Whereas it could rationally find a role in each and every volume.) And when one checks the two page references, he blinks in disbelief: it is but a name on two charts which schematize the branches of the various sciences. In one chart, it is named as a recent branch of biology. In the other chart which summarizes the preceding seven charts, the word ecology is not even maintained as a major branch of the life sciences, but it is named as a field that is interdisciplinary between the life sciences and the earth sciences.   One is pleased to find that

"Conservation and Natural Resources" is one of the 41 major headings in the index-chart of the final Guide, yet the inclusion of this heading seems almost like an afterthought. Three of the four references are technology-oriented, and the fourth quite validly refers to Luna B. Leopold's entire book on "Water." Yet "Water" is none the less a treatment of but one resource. One looks in vain for reasonable discussions of forestry, range management, wildlife management, none of which is even indexed.

One striking "indicator" of the ecologic sophistication of a publication is the treatment accorded to Malthus. In all 26 volumes, we find Malthus mentioned but once, and then he is but granted a place in the margin of a text page, where he is referred to as a "prophet of doom." With the customary short-sighted optimism of the farm technologists who are particularly devoid of ecologic thinking, we learn that "advanced nations, benefiting from developments Malthus did not appreciate, such as birth control and farm technology, produce more food than they can use."   Another test of the ecosystem sophistication of any publication is the recognition given to DDT, most insidious polluter of the Total Environment. We find it indexed as being in two volumes. In one it is praised for its control of malaria, with only a mention that the mosquitoes are becoming pesticide-resistant. In the other, concern is expressed for the manner in which DDT enters into and pollutes the Total Environment. Rachel Carson's *Silent Spring* gets attention; then her message is counteracted by a reference to the (obvious) benefits of pesticides in general (not just the persistent eco-system-poisoning chlorinated hydrocarbons). But the Life book was published in 1965, and I doubt if its author would not now write more strongly. René Dubos' volume on "Health and Disease," and Luna B. Leopold's "Water" are most commendable outstanding exceptions to the generalizations above. But then, such are René Dubos and Luna Leopold. It is regrettable that their knowledge and philosophy did not extend throughout the series.

As measured against the chart of the sciences used in this volume (p. 127) it may be said that the Life Science series essentially does not recognize the sixth, seventh, eighth and ninth Levels of Integration of the sciences. In terms of biology (p. 119) it is high on (Blue Version) molecular biology, but grossly non-cognizant of (Green Version) ecological biology. It looks sidewise to technology, to the "better life" of affluence and gadgetry for ourselves (and to hell with our grandchildren); it does not look forward with imaginative vision to a future of quality for mankind and his environment. In these respects, the Life Science Library was not only out of date before it was published, but by failing to communicate the higher message of these sciences, has probably retarded their growth at a very time in the history of civilization when they are most grievously needed. May the future additions and revisions of this series compensate for this sociologically unfortunate oversight.

LIFE NATURE LIBRARY

Life Nature Library (the word life refers to *Life Magazine*) comprises the following 25 titles: The Universe, The Earth, Guide and Index; Plants, Insects, Fishes, Reptiles, Birds, Mammals, The Primates, Early Man, Animal Behavior, Evolution; Ecology; The Poles, The Desert, The Mountains, The Forest, The Sea; North America, South America, Eurasia, Tropical Asia, Africa, and Australia.

New York: Time Inc. About 200 pages each. 1961-1965

This series, sister to the preceding Life Science Library, joins with it as being unquestionably the most prestigious and influential single force in publishing history for the popularization of its subject matter. I would smile, vehemently, with its title. The title beautifully reflects the semantic confusion in the public mind between science and nature, and because of the influence of this series, will probably fertilize and establish that confusion in the English language. As applied science (technology) is called "science" in the other Life series, so basic science (essentially inapplicable, in this age) is merely called "nature." One catches the insinuated difference: one is useful, the other is useless. One is practical, leading to a higher standard of living and a greater Gross National Product. The other is for maladjusted little ladies, and their male equivalents, with bird glasses and butterfly nets. How soon will we learn that b o t h are science, a single to-be-integrated body of knowledge about man and the environment around us?

There are no Consulting Editors for this series, as there are for the Science Library. The plan and format of each of the 25 volumes however is equally superb. Each volume, with few exceptions, is composed of eight textual chapters, each with a following supplementary independent yet related "picture essay." The textual chapters, with one exception, are by a single author, a scientist or a writer of solidly established reputation. The picture essays are by the "Editors of Life" and are predominantly photographs. I continue to feel that this separation is confusing to the reader. He is really progressing thru two books covering the same ground, while he is forced to alternate from one to the other. On the other hand, I suppose this duality does allow the solo author relative autonomy in his writing, while the hosts of editors can expend their energies on the picture essays. The fact that chapter titles are written in "literary English" that smack of the cute and the poetic, rather than with definite factual scientific words, leaves me more than annoyed. Subtitles in "scientific English" would not leave one guessing.

There is a worse fault in the editorial handling of the textual chapters. They lack subdivision headings (such as are in the Life Science Library). Altho each such chapter is well organized into 6-10 sections, each revolving around a specific topic, these sections are only recognized by a "paragraph initial," which itself is totally uninformative. I am sure that the editor who made this unintelligent decision never reads his own books in order to learn. For those who like to know what water they are jumping in to, who like to scan first, then read, then wish to recheck some back paragraphs, this design is an abomination. I would recommend that you write in your own subtitles—for when your friends read these books, or when you re-read them.

For the students who wish to give themselves a self-education with these volumes—and I can think of none better—I suggest the following sequence. I would start with The Universe, dealing with astronomy, the most inclusive science of them all. Then I would take The Earth, for geology is the science of the one heavenly body that is our own finite spaceship. Thirdly I would recommend the Guide and Index, which despite its title, is an excellent taxonomic survey of all the kinds of plants and animals and minerals which are the c o m p o n e n t s of our world. The rest of the volumes segregate into three series.

The *first* series surveys Fifth Level biology (see p. 124), with volumes on plants (botany), insects (entomology), fishes (piscology), reptiles (herpetology), birds (ornithology), mammals (mammalogy), the primates (primatology), and early man (paleoanthropology). But one volume on botany! In addition there is a volume on animal behavior (ethology) that bridges the first six of those eight volumes, and one on evolution (genetics) that bridges the first seven of the eight. Both second and third series can be said to be introduced by a volume on ecology which is an excellent treatment of the subject at the Seventh Level of plant- and animal-communities.

The *second* series of five volumes breaks down the earth into five biomes at the Eighth Level, that is, ecosystems of plants-animals-and-the-physical-environment. These five are: poles, desert, mountains, forest, and sea. It is extremely odd that the grasslands were ignored, for they certainly are a sub-world of coordinate rank to these others.

The *third* series divides the world from a chorologic or geographic point-of-view (see p. 133). There are six volumes, on North America, South America, Eurasia, Tropical Asia, Africa, and Australia, each one preceded by the common subtitle "The Land and Wildlife of." These are Eighth Level treatments that emphasize the differences and uniqueness of each continent.

Altho I have emphasized the distinction between the Science Library and the Nature Library as being one of the practical technologist vs. the basic scientist—and I think this distinction is reasonable—it is most interesting to see how the psychology of the authors, and the psychology of our American Way of Life, does show thru. Sometimes it is a matter of deploring the destruction by man. More often, the technologist-just-under-the-surface shows thru, as he twists to extract more "resources" for

more people, today, and to hell with tomorrow. The titles of
the l a s t chapters reveal much of our attitudes towards man
and the world about us. The geologist closes his book with a
discussion of "An Uncertain Destiny"; the herpetologist with
"A Dubious Future." As man's bulldozers keep on rolling, the
African wonders about "Wilderness for the Future," and the
Eurasian, with "The Last Strongholds." More often we find a
polarization and arrogance: The ecologist concludes with "Man
versus Nature," while the botanist is more certain, ending with
"Man the Master." In support, the piscologist writes of "The
Costly Struggle to Harvest the Sea," the polar expert of "The
Coming Boom in the Arctic," the desert scientist boasts of "The
Desert Tamed," and the forester sings paeans to the super-pro-
ductive "Forests of the Future." Only one author, the ornitholo-
gist refers to the stand that should have been the dominant
theme of all 51 books, "Toward a Balance with Man." The editors of Life, not content with this vision, titled the correlative
essay "Man, the Admiring Enemy." It is coincidence that the
51st book I myself read was "Early Man." The penultimate
chapter of this book is "The Dawn of Modern Man," and the
final chapter is candidly titled "The Persistent Savage."

Here is SCIENCE of the 1960's. Nowhere in all these volumes
do I find other than the faintest glimmers of a Total Ecology
concerned with man-and-his-total environment, the one science
directed to the continuance of men of quality, in an environment
of quality.

5. Methodology and Psychology

Scientists will always tend to forget they are merely human (that is, they and their immediate confrères—not their scientific adversaries). They would like to (as at times I would myself) act as a completely unemotional a-human not-light-restricted camera, capable of recording truth and reality.

Psychology, as the science of the mind, is one of the more sobering influences on this vanity of the scientist. Psychiatry is often the more relevant. In this section, we will sketch one sequence from among many, of psychologic phenomena that take place in all methodologic procedures. An understanding of the nature of its three stages will help us to realize the limitations of the empiricism that is basic to all scientific knowledge. Then we will sketch six compulsive quests that are part of the innate thinking pattern of human beings, and that will perennially arise as long as man is man, to distract him from his major quest: knowledge.

On from the Senses

> *Without all doubt, the fumes of faction not only disturb the faculty of reason, but also pervert the organs of sense.* —Tobias Smollett (1721-1771)
> *Humphrey Clinker*

1. *PERCEPTION.* Our first, and essentially our only possible, awareness of the outside world is thru our five accepted senses: seeing, hearing, smelling, tasting, and touching. Other "senses" are still in the realm of scientific controversy. We see by means of light radiations affecting the nerve endings of the eyes. We hear thru sound vibrations of the air affecting nerve endings in the ear. We smell when gaseous particles entering the nose or the mouth affect nerve endings in the nasal passages. We taste when small material particles affect nerve endings on the tongue. We feel a touch when there are contacts with nerve endings located on most of the surface of the body, as a feather on a foot, or the contact of a lip. In the first analysis, science depends on nerve-endings, and the person who is not adequately nerv-ous is no scientist, tho nervousness does not by itself make a scientist. This first stage, reactions to the effects on the nerve-endings, is known as "perception."

Actually, our senses are by no means at the pinnacle of biologic evolution. Other animals have been treated much better by their gods. Eagles see much better. Various animals "see" ultra-violet and infra-red. Dogs and insects smell much better. Many a male moth will catch the fragrance of his female at a distance of several miles, and will zero in upon her (to find other males there already), while woman must resort to the artificiality of a perfume. Rabbits hear much better. Bats are capable of supersonic echo perception. Bees perceive the plane of polarization of light from different parts of the sky and thus orient their flights. Certain west African fishes send out and receive electrical impulses. Birds can apparently use the sun as a compass, correcting for change of solar azimuth thru the day. It is entirely possible that animals have sense perceptors making "known" to them aspects of the external world far beyond our present comprehension.

Furthermore, our senses are not entirely reliable, or what we think are our senses. Ghosts, mermaids and sea-serpents, or their equivalents from outer space, even make their way into the respectable scientific literature. Messages from the distant and the dead have come to us with monotonous regularity thru the ages, to be believed by sages and sinners.

2. *COMPREHENSION.* Under this term, we can loosely aggregate a variety of psychologic phenomena, all dealing with what happens to those sense impressions after they have been perceived by the mind. What the mind does to them, or how it does what it does, we know relatively little about. This itself is a field of scientific activity which future centuries may open.

Comprehension is linked to "intelligence" which, though we have many measures of it, we know relatively little of. It is tied not only to one's inheritance but also to one's cultural background and conditioning, and one's preparedness to understand these sensations. What we have already discussed as "concepts" and "logic" now find their place as part of this general psychologic field of comprehension.

> *I always accounted as extraordinarily foolish those who would make human comprehension the measure of what Nature has the power or knowledge to effect; whereas, on the contrary, not even the least effect in Nature can be fully understood by the most intelligent minds in the world.*
>
> —GALILEO (1564-1642)

3. *EXPRESSION.* Perception and comprehension are not adequately sufficient either to ourselves or to society, unless we can get these facts and ideas across to where others can receive them. Thus, for want of a better term, "expression" becomes the third and final stage of this psychologic sequence. Expression is the world of speech and writing, the world of words, numbers, symbols, figures, charts, pictures and graphs, the world of linguistics. Do not think it is a perfect world. Whatever troubles you have in reading this book—or I had in writing it—are parts of that imperfection. The same word is often used for many different concepts. The same concept is called by many different words. Confusion and more confusion! The entire field of linguistics, in one sense, can only be in terms of symbols of the known. Thus the saying of something really new is fraught with almost insuperable problems. There is many a slip twixt one mind and another.

REVELATION. It may not be out of place at this time to mention the role of revelation as a psychologic phenomenon. Here is one source of knowledge, as claimed by a large segment of humanity, that comes not by the known senses from without, but by revelation from "within." In this category, we have much of the occult, the mystic, and the spiritual. We have the experiences of the great religious teachers of all history. We have the intuitive, the purely logical, and the apriori.

From this enormous melange come not only the pseudosciences and the -mancies, cultism, fraud and quackery, but also the great religious scriptures of all time, and even some of the finest of scientific insights and conceptions. The human mind is the deepest of unplumbed wells. Psychologists and behavioral scientists must always show respect for revelation. It is an uplifting and enduring expression of the human need to master the inexplicable.

Six Quaint and Queasy Quests

The leading inclinations set the plots. First is credulity, *the will to believe —to believe not wisely but too well. Second is the thrill of* marvel, *which is bound up with magic. Magic is the primitive explanation of how things happen; it is the child's favorite fiat, for there is no need for a deeper philosophy. Against this Weltanschauung, which begins in the nursery and continues by right of primogeniture, the recognition of the true order of nature had to make its way. Close of kin is* transcendence, *the crediting and claim of supernatural powers; and everywhere, once committed by whatever route, the* prepossessed *mind finds what it looks for. The beliefs chosen among the many called owe their preferment to congeniality of conclusion. One and another individual solution, when pursued with fanatical singleness of mind, becomes a* vagary; *while in all the higher reaches of thought there is an appeal to evidence, a form of attempted* rationalization, *which is the pride of the intellect. Neither deadly sins nor desirable practices, the* seven inclinations *form deviations from the path of wisdom by yielding to wish.*

JOSEPH JASTROW
Error and Eccentricity in Human Belief. 1935.

Psychology plays many roles in science and in scientific methodology. If we are to accept science for what it is, then we shall not only think of it as man's noblest intellectual creation, but a creation shot thru and thru with many unarticulated major premises that are part and parcel of our compulsive emotional nature. Wishful wisdom is the way—that seduces man—unless woman is around.

1. The Quest for *CERTAINTY*. There is nothing more satisfying than being sure. Much of our scientific activity is related to this quest for certainty, to unveil the mysteries, to clear away the doubts, to leave no stone unturned, to "know" in a completely emotional sense of the term. There is nothing wrong with this quest—except that we should never forget that it is the quest which leads to science, not to the attainment thereof. Beware of certainty.

2. The Quest for *ORDER*. There is nothing more satisfying than unfolding the law and order in the universe. Our minds seem to want this order, this system; and so strongly do we want it that we often find it, even when it is not there. There is nothing wrong with this quest—except that we should never forget that we are thus seeking to find—what may not be there. Beware of order.

3. The Quest for *TRUTH AND REALITY*. There is nothing more satisfying than attaining ultimate truth and reality. Mankind has searched for it as long as he has thought. There is a strong paradox here. These two concepts hardly belong to science, and they are ones which will give us considerable trouble. Most laymen consider that science tells us what is "true" and what is "real." There are many scientists who believe this also, and who will stoutly defend the "facts" of life in this light. Nevertheless, one has only to read the history of science to realize that scientific "truth" and scientific "reality" are constantly changing as our knowledge changes and develops. We must be ready to grow with science, otherwise we become one of those "fossils" known so well to college students. In truth however, if you want an ultimate "truth," a final "reality," it is fundamental religion, not science, that can give it to you. It is faith, not doubt, that creates it. The wisdom of science lies in the admission of ignorance and uncertainty, uncongenial bedfellows for the soothing seductiveness of truth and reality. Beware of seductiveness.

4. The Quest for *CAUSES*. There is nothing more satisfying than knowing what "causes" things. The search for "causes" is unquestionably the most compulsive single factor that leads scientists to play the games they do. There is nothing wrong with this quest—providing it does not lead us to the end of the road. If our search for causes does not raise more questions that it answers, we are laymen, not scientists. Beware of causes.

5. The Quest for the *BALANCE OF NATURE*. Every race of mankind, every individual human being, weary of the strife and turmoil that in evolution has made Homo sapiens what he is, has sought refuge in a place of peace and quiet, where the lion laves the lamb. (No normal lamb, that!) This is the Garden of Eden, the Golden Ages of historians of every age, the "climax" of American ecologists, the "peneplain" of geologists, the "steady state" of a "closed system" of other scientists, the "ideal marriage" of starry-eyed young hopefuls.

There is nothing wrong with this quest—if we do not forget its emotional basis, and that what we see in nature is a trend towards an equilibrium—but an equilibrium which is never attained, or attainable. Beware of the balance of nature.

> *In general let every student of nature take this as a rule —that whatever his mind seizes and dwells on with peculiar satisfaction is to be held in suspicion; and that so much the more care is to be taken, in dealing with such questions, to keep the understanding even and clear.*
>
> —FRANCIS BACON (1561-1626)
> *Novum Organum*

6. The Quest for *QUANTIFICATION*. When a biologist—who deals with the larger holistic units of the world—runs out of ideas and does not know what to do, he begins to count things. He thereby steps through Alice's Looking Glass, into a wonder-world of numbers. It is a landscape where, with every glance, one finds mincing mathematical methodologies waiting to be picked up, walked with, ridden, even at a gallop. The young quantifier, basking in the aura, has stars in his eyes. He is on the high road to attaining sublime truths, which lie just around the next curve, and which in the meantime mystify and impress the nonquantifier that has been left behind. These opinions are no belittlement of mathematics, assuredly the queen of man's intellectual castle. Too often, however, one who grasps the magnificence of mathematics either cannot write English, or knows that English is no substitute for mathematical symbolism; and even he who does not comprehend mathematics can ape the talk, as can a trained parrot. Too often a biologist adopts a mathematical method with no more comprehension of its values, and its limitations, than the computing machine into which he feeds the data. The quest for quantification should be questioned most seriously before the quantification is accepted as the sole soul-goal of the quest. Beware of quantification.

There are many other quaint and queasy quests. They are reasonable spurs and drives to sound scientific investigation. When they dominate however, as they sometimes do even in the pages of our most prestigious scientific journals, we have a world of wishful wisdom, an era of error and eccentricity in human belief. This is SCIENTISM, not science. It is the "science" of the layman who loves to be so laid— infertile futile fun.

6.   Methodology and Instrumentation

*We see instruments turning from servants into tyrants, forcing the captive
scientist to mass-produce and market senseless data beyond the point of
conceivable usefulness—a modern version of the Sorcerer's Apprentice.*

—PAUL WEISS. 1962

An instrument is a contrivance, a tool, a utensil, an implement,
which allows, improves, or extends any activity that would otherwise
use but mind or muscle. Cain slew Abel more efficiently with the jaw-
bone of an ass than with his own fist, altho many varieties of other
animals before him used tools for food and foe. Since then, men and
asses have contributed to a truly stupendous proliferation of the field
of instrumentation. For example, the "1969-70 Guide to Scientific Instruments" of the journal *Science* (September 23, 1969) is essentially
a 154-page fine-print index, plus 194 pages of advertisements. Dazed
with this brightness of our technology, I wonder if some are not inclined to forget that the most important instrument in science must
always be the mind of man.

Instrumentation is here viewed as the abstract consideration of instruments. We should never forget that modern science could never
have developed without the concomitant developments of its instruments. It is well, therefore, to give them a moment of separate consideration, for in many fields, methods are only as powerful and productive as are the instruments they employ.

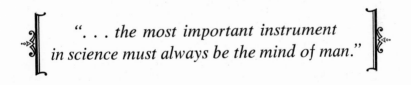

*". . . the most important instrument
in science must always be the mind of man."*

We may consider the nature of instruments in three basic categories:

A. *Extension of Sensation.* On previous pages we indicated the range
and limitation of the five human senses. When our sense organs won't
perceive certain phenomena, the ingenuity of man has created instruments for the purpose. In this manner, we have not only extraor-
dinarily enlarged the range of our sensations, but we are perceptive
of phenomena quite unrelated to our sense organs. The narrow band
of visible light radiation has been enlarged in one direction thru
ultra-violet, and in the other direction past infra-red and out thru

radio waves. Audible sound has become part of a vibrational sequence that includes the supersonic. Magnetic and electrical phenomena are perceived now as common every-day experience. The microscope has revealed a world of the minute. Aerial photography has brought us more than a bird's-eye view. The telescope has revealed a world of the huge and the distant.

B. *Computation.* As scientific data accumulate by direct instrumentation, they need to be organized, analyzed, and worked over by computational processes. The development of complex instruments for these purposes has not only allowed some procedures to progress more quickly and with less possibility of error, but has actually initiated new computational procedures. One of its simplest forms is the adding machine. At the other end of the scale are the electronic monsters with built-in "memories," that run through computations in a matter of hours that previously would have taken life-times. As a result there are advances in science of which we would not have dreamed before the advent of these machines.

"But don't let anybody fool you," said Monroe Rathbone, chairman of Standard Oil Company, New Jersey, in 1965. "No computer can think, and none ever will. In order to use the computer right we must be more sure than ever that we have analyzed our problems correctly. It might be a good idea to paste on the front of every computer that old saying: 'Ask me a foolish question and I'll give you a foolish answer.' "

C. *Automation.* The subsequent development in the field of instrumentation concerns the automatic control of complex operations, themselves requiring a variety of instruments. These are essentially independent "systems," with feed-back mechanisms to allow for self-correcting operations, which in turn keep the "organism" going even tho there are alterations within itself, or in the external environment. In business and industry, automation at the beginning of the second half of the 20th century was being looked upon with mingled fear and awe. These instruments are taking over routine work loads and automating the activities of what had previously been accomplished by hundreds of laborers and technicians. Certainly these developments will radically change many phases of human society. What it may do in science is still over the horizon. Whatever it is, will be phenomenal, for the instruments of the future have even greater promise. Already, the non-manned orbiting satellite provides a veritable separate world of complex instrumentation that symbolizes more than any other single object the cleverness of man.

*   *   *   *   *   *

There is unfortunately a strong emotional compulsion to synonymize instrumentation w i t h science. In an affluent society, instruments are "things," that impress other people. The corporations producing them help to create their own markets by convincing scientists that they are not scientists unless they do have this expensive gadgetry. In turn, science attracts the instrumentophilous personality-type. And thus this biologic ecosystem snowballs, until it disparages and ridicules the scientist who cannot display a laboratory loaded with instruments. Even in 1967 a statement was made at the huge national meetings of the American Association for the Advancement of Science that "History is sometimes represented as the most useless of disciplines on grounds that it cannot be used instrumentally." Taxonomy, as a scientific discipline, was essentially dead in the mid-century. It works with actual plants and animals; not instruments. The only live twiggies of the field are those that have gone to cytology and genetics (and use a microscope) , and to "mathematical taxonomy" (and spend their time counting, to use a computer) . The entire field of biology has swung leftward to "molecular biology." It offers unlimited opportunities for instrumentation. I recall the professor of a well-known woman's college telling me that the administration was insisting she buy "equipment." She really did not need any. A computer would do; it was not specific, and could be used for a variety of purposes. Now she has it. "But Frank, what d o you do with it?" she pleaded. I am not sure what Darwin would have done with instruments, instead of bones. There never were enough bones anyway, to interest the instruments. We probably would not even have had Darwin, if he was trained to love instruments instead of bones. And yet, we still need Darwins in science.

Man is the measure of all things.

—PROTAGORAS. (c. 481-411 B.C.)
(quoted by Plato, in Theaetetus)

7. Methodology: Observation versus Experimentation

The observer listens to nature; the experimenter questions and forces her to unveil herself.
—Baron Georges L. C. F. D. Cuvier (1769-1832)

1. The Observational Method

Earliest in terms of historical development, still important, and probably always to be important, is the so-called Observational Method in the study of natural phenomena.

Under this term, we include all those studies which are concerned with natural or semi-natural conditions, not purposely modified by man for the sake of the investigation. Such scientific work is carried on largely by extensive reconnaissance over a wide territory, rather than by detailed local studies; or thru many years, rather than in a short period of time. The data are more often, but not necessarily, of a qualitative nature, and are said to be "subjective" rather than "objective."

It was the Observational Method that was used by all the early natural scientists. By the mid-20th century, however, it was the conventional wisdom to belittle such methods as old-fashioned, inadequate, and even "unscientific." That the method can be abused and has been abused, is true; but that is so of every method. Furthermore, when a bright young scientist with money to spend, a paper to write and publish, and a degree to get, all on a short-order basis, when he goes into his lab or into the field equipped with a carload of costly instruments, that will yield data demanding higher mathematical analyses, then by comparison the observational scientist who wanders and putters around with nothing but a notebook and pencil and maybe a camera, and possibly doing more sitting-looking-thinking than wandering, he makes a poor showing indeed. Nevertheless, the observational scientist at his best has one tool, one germinal creative plaything, which the other need not have and often does not have, and which puts the observationalist at a greatly superior advantage. That tool is the human mind.

Observational and interpretational methods need not be restricted in points of time and space to presently existing conditions. Repeated observations thru months and years can yield magnificent data on change, involving alterations entirely unanticipated at the start of the study. Furthermore, altho a sharp scientist may very thoroly have investigated ten valleys and feel fully justified in drawing authoritative conclusions, yet a trip to the eleventh valley, however brief and hurried, may provide the key information to unlock a magnificent scientific discovery that revolutionizes the interpretation of all the other valleys. So it has been repeatedly, in the history of science.

One of the great values of a high-quality observational scientist is that he can go into a new region or field, and in one relatively short period of continuous study, be it for a day or a month or a year, come up with a sound interpretation of the natural conditions, involving the history of its past, and predictions as to its future. Indeed, when time factors are greater than human lifetimes, at least greater than the duration of the research grant, then sound interpretive methods epitomize the brilliant scientific intellect. He is followed by the hordes of aides and technicians who, twittering like starlings in a roost, eventually prove to their own satisfaction that the scientist was right in the first place.

> *To the eyes of the man of imagination,*
> *nature is imagination itself.*
> —WILLIAM BLAKE (1757-1827)

2. The Experimental Method

Without imagination one can contrive infinite variations of experimental set-ups, all of them novel, yet utterly uninteresting, inconsequential, insignificant. The mere fact that something has not been done or tried before is not sufficient reason for doing or trying it.

> —PAUL WEISS
> *Experience and experiment in biology. 1962*

Man loves to fiddle with things. He manipulates, alters, controls, and gloats in his sense of power, not content to leave things be. This not particularly inspiring attribute of the common man has been turned to very creditable advantage—and often to very discreditable disadvantages—in the highbrow art of experimentation. An experiment is a refinement on observation, in which phenomena at a particular time and place are carefully watched in relation to artificial changes that occur in them or around them. Basically, an experiment involves an intentional disturbance: thereafter it is really no different than the observational method.

A "controlled experiment" is one where two or more such situations are watched, where one is observed without interference, and where the others are manipulated in known ways. The changes are then related to this known manipulation. The "designed experiment" is a term for one that is set up with particular reference to the resulting data that will be statistically analyzed according to a plan or design. Such experiments of course, are no better than the subjective judgments that were part of or preceded the actual design.

Thanks to experimentation, our scientific knowledge of the world about us is vastly greater than it otherwise could be. Even tho this is more true of the physical sciences than of the ecological fields, the experimental approach to the natural sciences must be an integral part of the thinking of every researcher.

Claude Bernard

AN INTRODUCTION TO THE STUDY OF
EXPERIMENTAL MEDICINE

With a biographical sketch by Paul Bert, 1878. English translation by
Henry Copley Greene, with an introduction by L. J. Henderson, 1927;
reprinted 1949.

New York: Dover, reprinted with a foreward by
I. Bernard Cohen. 226 pp. 1957

A mongst the books reviewed in these pages, I recommend
none more warmly than this, which might the better have
been titled "The Philosophy of Experimentation, as exemplified
by Medical Science." This is a timeless and priceless gem that
will take its place with the classics of antiquity. Dr. Bernard was
firstly a scientist, secondly a biologist, and thirdly a medical
scientist. He writes with candour, honesty and simplicity, fully
aware of the fables and foibles that dominate his field and the
men in it, yet most highly motivated to improve that field. His
was a spirit so rare that his light is continuing to shine long after
his death.

The book finds its logical place in this section on "Observation and Experimentation" and not as coextensive with a total
philosophy of science. During Claude Bernard's lifetime, medicine was an empirical rule-of-thumb practice, in which the
physician could always bury his mistakes—a fine art not yet
dead today. Experimentation leads the way to a rational medical
science. One can comprehend therefore not only his insistence on
Determinism as being almost synonymous with Science (for
without known causes leading predictably to stated effects, a
physician would only be a gambler at the race-track). One can
also understand his disregard for statistics (for the actuarial
tables of the insurance companies, tho sound ethology, leave no
love for the life of the individual man).

The book is logically divided into three sections. The first,
on theory, is entitled "Experimental Reasoning." The second,
on basic science, is "Experimentation with Living Beings." The
third, on applied science, is "Applications of the Experimental
Method to the Study of Vital Phenomena," of which the last
chapter "Philosophic Obstacles Encountered by Experimental
Medicine" is an elaboration upon the "obstacles . . . arising
from vicious methods, bad mental habits and certain false ideas."
He is modest, and realistic: "I am only trying to guide men's
minds toward a goal . . . Time will do the rest. Of course, we
shall not see scientific medicine blossoming in our day, but that
is man's lot; those who sow and laboriously cultivate the field
of science are not also destined to reap the harvest."

8. Methodology and Sampling

The *philosophy* of sampling, and the *methods* of sampling (or "sampling procedures") are two very different frameworks of thinking and acting. All too often certain methods of sampling will be applied willy-nilly, by some unthinking technician or aide with little or no consideration that a philosophy is also involved. Yet it is the underlying philosophy which conditions or even determines the logicality, or the ludicrous nonsense, of the results.

The philosophy of sampling, the idea of sampling, in its simplest form is based on a situation in which the whole is too great to be known in detail. Note that "the whole" is already a clearly formulated and agreed-upon concept in the mind of the investigator, whether it is a roomful of air, a cabbage patch, a river of water, a spruce forest, or an urban ghetto. This is the first hidden assumption in the methodology. The whole can be entirely subjectively determined, the choice of emotion bias and prejudice, even tho the subsequent objective methodologies exercised upon it may give the study an aura of the purest intellectual sanctity.

Having chosen one's "whole," one chooses parts or "samples" which are to be considered representative of that whole. There are many many ways of choosing samples. They vary from being openly and candidly subjective (which means they are as good as the brain of the observer, fine in the case of a master mind but rather awkward when one has to employ aides and is scraping the bottom of the barrel to get them) to being purportedly objective (which means they are flawlessly logical, according to some ideal mathematical principle). These samples are then studied in all the voluminous detail that the research program permits. Then the scientist generalizes *from* his concrete knowledge of the parts *to* an abstract knowledge of the whole.

I am sure that an intelligent but untrained youth can see that no amount of rigorous and excessively detailed study-of-the-parts can make the total study any less whimsical if the p a r t s were chosen in a whimsical fashion, or if the w h o l e was chosen in a whimsical fashion. And if both were chosen by whimsy, the study itself is purest whimsy. As one reads this scientific literature, it is a bit disconcerting to search and search the pages, and find no indication whatever as to whether whimsy or wisdom determined the choice of parts and wholes. One must have pure faith in the author, even as with the high priests of old.

Sampling of crowd-phenomena may reasonably be accepted as the sampling of acknowledged homogeneous materials. In this category are such things as the billions of molecules in a gas, blown sand from river and shore dunes, and river water. We might include colonies of individual bees, and crops of uniform agricultural lands. We also have mechanically homogenized materials, like wheat flour from bins, packaged foods from a grocery store, and sausage meat. It is rather interesting to note that when the scientists turn up differences in measurements from replicated samples of such assumedly homogeneous wholes, theory demands that the differences are due to poor sampling techniques (not to real variations within the whole). Thereupon it is the methodologic techniques which are improved. Another hidden subjective assumption!

These are the methods, fine as they are for homogeneous materials, highly developed in their rigid mathematical and quantitative aspects, which have put stars in the eyes of those scientists who are involved with the higher levels of integration of ecological systems. These ecologists m u s t ape the manners and morals of the molecular men, for such *is* the science of its age. They are actually on the horns of a dilemma. As one studies higher and higher levels of integration, involving larger and larger wholes, one eventually finds himself concerned with *the* Earth. There is only *one* Earth. Thus, a "controlled experiment" is impossible, for there is no second Earth to use for comparison. And to "experiment" with the one Earth is something no organism claiming intelligence would ever do. Yet politicians, the military, and the technologists have arrogated unto themselves the right to so experiment—as in high-level atomic blasts that have already affected the Van Allen radiation belts surrounding the Earth, involving hazards that will lead historians of the future, if there are any, to describe the stupidity of the savages that then inhabited the planet, and the strange religion that then dominated the thinking that led to such self-destructive practices, practices that were but one step above those of the Mayan priests who lived but a few centuries earlier, and tore the bleeding hearts from the chests of their living sacrifices.

Nevertheless, scientists dealing with the larger ecological systems have solved their problems in several ways that at least are satisfying to themselves. In a sense, these are tricky intellectual gymnastics for which psychologists must have a certain amount of admiration. They really do not solve the basic problem, but they serve as a sort of "substitute behavior pattern" which gives these people satisfactory, even if sham, status within the ideological liturgy of the times.

In the first case, the classical biologists—those dealing with the species of animals and plants—have almost wholly succumbed to "molecular biology" and its magnificent minutiae. I do not belittle the field. It has revealed extremely interesting and significant information about ourselves and other organisms, even tho in that process our universities have lost their Botany Departments and Zoology Departments, and the new "Biology Departments" (sic) are all but bedroom denizens of the virile chemists and physicists. The simple trick here was to focus not on separate organisms (not enough of them), nor on organs or tissues (not enough of them), nor even on cells (not enough of them), but on the chemical molecules composing these organisms. Presto! The biologist had crowd-phenomena to study. He could again become a "reputable" scientist.

The larger ecological ecosystems took a bit more fancy finessing before they would fit into the reigning dogma. But it is being done. "Population ecology," or "ethology," involves herds, flocks, and large groups of individuals of one species. Here mathematics came to the rescue. Altho populations of insects provide the larger, and thus more satisfying, numbers game, even populations of larger animals are amenable to the mathematical methods. The highest praise of one scientist to another in this field is to say of him, "He is highly mathematical. He is highly theoretical." (Catch that sanctimonious tone of voice?) With large numbers, one can sample populations. The problem is solved.

When it comes to bio-communities (composed of several different species) and natural ecosystems (composed of organisms, soil, air, water, plus other unrecognized elements) there is a new twist in the matter of sampling methods. Samples (from an assumed homogenous whole) are now chosen and studied. Differences amongst the samples again show up. Previously, the differences were assumed to be due to poor techniques, and the techniques were "improved" (so that significant differences did *not* show up!) However, the differences are now assumed to be due to differences in variables that went unrecognized in the original sampling—a radical and significant change in underlying philosophy! So what do we do with the differences? We interpret them in terms of the "probability" that we really have samples of *more than one* "homogeneous whole!" There follows some magnificent juggling, and we come up, not with several accepted distinct homogeneous wholes, but only with the tantalizing "probability" that such different wholes do exist. In other words, the variable non-homogeneity of the world about us can be scientifically sampled and studied only within the philosophic framework of an assumed homogeneity— hardly a satisfying intellectual pastime—except for the scientists involved, and for their lay congregations who "believe" in what they are doing, and pay for their instruments and their salaries.

To force the sampling of the larger natural ecosystems into a crowd pattern that would minimize this nasty little habit of having differences show up in the sampling that would imply the probability of unrecognized variables (a slur on the scientist) and of more than one homogeneous whole, these scientists broke their wholes down into smaller and smaller bits. They would never admit, but any fool can see, that the smaller the particles of mince meat, the more one can ignore and forget the larger components, even if widely divergent, that originally went into the grinder.

For example, how would a scientist "sample" a monkey, one single, whole monkey, that surely acts as a "whole" in nature? He could follow it thru from birth to death, as the oldtime family physicians used to do. He could consider it as one individual in a tribe of monkeys. But in each case, voluminous mathematical data, such as our computers demand, would not be available. On the other hand, he could freeze the monkey. Slice it, like a loaf of bread. Take each slice, thoroly macerate it, mix it all up, homogenize it so it c a n be sampled. Then analyze samples of one slice for their chemical constituents. So treat each slice. The head-slices with their gray matter would be thrown out as "atypical"—heads always cause trouble. The resultant mountains of monkey data could be treated mathematically in endless ways, and add many many pages to the scientific literature. The end results may not be a good monkey description, but I can vouch for the fact that a lot of such monkey business is going on amongst our philanthropic foundations, our federal granting agencies and the research factories we call universities.

To be more specific, such business in ecosystem science is currently riding the crest of an enormous wave, a tsunami that has literally flooded out and all but destroyed much valuable and needed scientific research. I refer to the fashionable trends of the 1960's for "productivity studies," "energy flow," and "nutrient cycling." These are all facets of research—fully important in their way, as facets of some larger total-research program—that satisfy the insatiable methodologic demands for short-term research, yielding vast quantities of data, obtained by grinding up the ecosystem into such minuscule particles that the Crowd Phenomena *are* there, *to be* sampled, even if the holistic nature of the monkey is lost in all the monkey business. Millions and millions of dollars are going into this type of research— itself a human ecosystem of its kind, so closely knit that the universities, the government agencies, the foundations, the administrators, the leading scientists, the graduate students, all mumble the same ritualistic abracadabra, and contrive to keep the system going—for which the general public pays. The clearest evidence of this social phenomenon lies in the publicized plans of the International Biological Program, the most prestigious and expensive scientific endeavor that biologists have ever mounted—footed mostly by the U.S.A. "Productivity" is the keynote for all parts of this program, with only the slightest lip service done for other and needed segments of ecosystem research. I see no hope for ameliorating the situation. Were the minced monkey scientists to be removed from the troop, the ranks would be so sorely depleted that not enough would remain to regroup and handle the millions of dollars that are being appropriated for this work. The productivists will surely produce. I only regret what other very important scientific research is remaining unproduced.

Characteristics of Samples

Since sampling is the taking of a part, as representative of the whole, much thought and experimentation have gone into several aspects of the problem that are here worthy of mention, especially for the larger samples used in the natural sciences.

1. *Size and Shape.* Almost every conceivable type of sample has been tried, and found useful for certain purposes. These are firstly *points,* used especially for fine-textured grasslands of interest to the range and pasture ecologist. *Lines* are used both in forest and grassland. *Squares, rectangles* and *circles* (all commonly miscalled *quadrats,* the word is impressive) are of various sizes, generally under 10 meters across. A *transect* is an elongate rectangle, often many kilometers long. There may be combinations of these samples, or samples within samples, or nested samples. The size and shape depends of course on the material one is sampling, and personal choice plays a part.

2. *Distribution of Samples.* There is much pious platitudinizing as to h o w the samples should be distributed thru the whole that is to be sampled. Some recommend that they be distributed *at random,* a thought that is mathematically satisfying, but awkward when one realizes that natural phenomena are not themselves distributed at random. Secondly, the samples may be spaced uniformly according to some predetermined pattern or geometric arrangement laid by precise measurement or roughly paced out. Those who like such order will find a reason to be geometrically oriented. Thirdly, there may be an arrangement in definite relation to some variation in the natural phenomena, such as from the middle of a lake to a cliff-top on the shore, or from the summit of a mountain to its base. Fourthly, samples may be placed arbitrarily, purposely, selectively and subjectively where the investigator chooses to place them. In the last analysis this is the most honest way, for in the other ways the subjectivity of the investigator is merely cloaked under the objectivity of a technique.

3. *The Number of Samples.* Many a scientific spotlight has been focussed on the problem of how many samples need be taken (thus avoiding time and effort on unnecessary sampling), but I am not sure how much light has been shed on the subject. The problem is often stated in terms of the percentage of the area sampled; and seems to be answered in terms of something that sounds like statistical adequacy. In short, when additional sampling does not essentially change the averages and means, etc. already obtained, you have already taken too many samples. (Of course ,the very next sample you might take could be the head of the monkey described above; but I am unfair.) We are back to that nagging thought that if the whole were not homogeneous—and you know it is not, because nature is not—you should not be "sampling" in the first place.

4. *The Homogeneity of the Samples.* Some scientists, seeking games to play and ways to keep busy, are devoting their energies to cogitating upon the significance of sample-homogeneity. In short, can they really study one slice of frozen monkey (and from it obtain help in understanding the whole monkey), or should they limit themselves to hunks of real bologna? Obviously, bologna is a far more suitable environment for designed, refined and sophisticated mathematical procedures. The last time I met with this problem I was asked to evaluate a proposed project that would cost high in six figures, clearly loaded in favor of a diet‚of bologna.

5. *The Testing of Sampling Methods.* Since sampling is a highly subjective and personal procedure, and many methods are available from which to choose in any one instance, the neophyte scientist is non-plussed as to which one of many is appropriate in a specific instance. Thus another bit of gamesmanship comes into being: comparative studies, where several sampling methods are applied to the same whole, and one decides which is "best" (still without knowing the whole whole). As I have seen these studies, if there are no "significant" differences between the data of the different methods (if there are differences, the implication is that the data come from *different* wholes; and since we know they don't, the philosophy is a bit absurd), both methods are equally valid (or equally invalid, a point which is not mentioned). Actually, unless one does study the entire whole (and he cannot, or he would not be sampling) there is no valid standard of comparison anyway.

And finally if one does completely convince himself that he does have a sound sampling scientific procedure for the whole being studied, some large ecosystemic unit, he must still admit that this is probably the only ecosystem of its kind in the world. The next such whole to be studied, be it a lake or an island, is probably sufficiently different so that he cannot rely upon applying all this refined sampling research. Unsettling, to the scientist who does want to think.

At this stage, an honest natural scientist may admit to himself that any elaborate sampling methodology is merely the justification and "proof" for hypotheses already reached by intuitive subjective and observational methods, and possibly unnecessary. He may restrict himself to a minimum of quantitative data; or as part of a government, industry or other organized team, where the data may be more important than the philosophy, he may go on and on and on and on. In any event, the natural sciences being what they are, the personal and subjective judgment of the researcher is still the foundation upon which the objective edifice is raised.

The non-sample is forever unknown.

9. Methodology: Averaging versus Ordination

The natural scientist, may we assume, has now decided on a sampling method, and on the size shape distribution number and homogeneity of his samples. He proceeds to take data from these samples. (All the rest of nature can now be ignored, and this is very comforting, for much of nature is simply not amenable to sampling procedures; it is too heterogeneous.) Sometimes the very quantity of his data overwhelms the scientist. He proceeds either to publish raw data, or bureaucratically files it way and earns his pay by accumulating even more data. It piles up, like garbage collecting on city streets during a garbagemen's strike, until some day it all gets tossed on the dump.

The organization and treatment of such sample data are coming more and more within the scope of biometrical and other mathematical concepts and methods, but even before these sophisticated tools are applied, the scientist must make a most critical and important decision, and a subjective decision regardless of how he rationalizes it into a false objectivity. This is a choice as clearcut as between left and right, as between black and white, as between good and evil. (And how evil is the choice he does not take, if we are to judge from the veiled emotionalism in the scientific literature!). There is no fence-straddling or happy medium here. It is one or the other. Will he adopt the concept of *averaging?* Or the concept of *ordination?* (Now averaging "proves" the homogeneity of nature, and ordination "proves" the heterogeneity of nature. It is, however, most unfair for me to mention this now. It is only the objective and unbiased computer which says one or the other at the end of an expensive and sophisticated experiment.) Sometimes a scientist likes to think that he adopts both concepts. Such a view is only because he is involved with two levels of spatial or temporal integration. For example, he might view his monkey as composed of separate and distinctive feet head tail and body, which can be ordinated according to differences in their chemical components; but each such unit is now a separate and distinct homogeneous mass of minced monkey meat. Carry thru this analogy to larger social units and ecosystems, and the strange parallel will be clear.

Scientists in many fields have battled with this problem, as long as there have been organized sciences. And the pendulum has swung one way after another, from generation to generation. Often it moves from the creative contributions of an eminent scientist, to the dogmatic disciple-professors of the next generation, to the rebellious students of the next, until a new creative scientist rises for the moment. Look at any map showing the ethnic origins of urban neighborhoods, or the forest types of a mountain range. You will see homogeneous blocks of territory, with fence-like lines between them as sharp as stone walls between pastures. Go in to such lands and—if you believed the map —you will be surprised that South Europeans do not live all on one side of a street, and North Europeans on the other, that Caucasians are not wholly in one block, and Asiatics in another. Geographers, ethnographers, sociologists, as well as soil scientists, meteorologists, botanists and zoologists have each had their problems. On the one hand, they see continuities, uniformities and homogeneities; and they describe them and map them. On the other hand, they see discontinuities, variations, and heterogeneities; and they describe them and try to map them. I always like to think that our emotional compulsions are on the side of the separatists. Each human being, each of us, has a sense of his own individuality and distinctness, even at the time of closest togetherness; and we keep looking for this in nature. We are not like a sponge, or a piece of sponge—for a sponge does not seem to care which it is—it lives along either way. The rest of nature, on the other hand, is composed of *merging phenomena,* of relative continuities, separated by relative discontinuities. This is very immoral for the human mind, but we struggle along for better or worse, sometimes *determined* to prove that the world is composed of separate things as distinct from each other as we are, and sometimes *determined* to prove that these are interdigitating interflowing parts of one whole, like the parts of our body. The problem is not with the world, but with our mind.

Averaging

The methodology of *averaging* has its roots in the compulsive emotional behavior of wanting to view the world as composed of distinct and separate entities, each relatively homogeneous within itself and yet very much like certain others. This philosophy allows for various orderly and hierarchic classifications, such as the form-species-genus-family-order system of the plant and animal taxonomists, who are certainly the most neatly orderly of all scientists. In this approach, the worker decides that all his data are from one "unit" of nature. He "objectively" determines this by subjectively deciding that some particular "classificatory factor" is all-important, such as skin-color, or a fold in the corner of the eye. He pours all his sample data into one bowl, mixes them around, describes this goulash, and ends up with "averaged" data, composed of averages, means, and other such arithmetical concepts. We are all familiar with these intellectual escapades. Our average income for the year, the average temperature for the month, the average number of children in a family, the average number of people in a square mile, are all parts of this approach to nature. By the act of averaging, we "prove" the existence of discrete and separate units in nature. The result is inherent in the method. The method is inherent in an emotional compulsion.

Before the scientist adopts the Average in his studies, he should think of the characteristics of the Average Nation, taken from data of the members of the United Nations. Such a nation has an average population, including an average number of negroes and whites. It has an average Gross National Product, an average income per person, an average number of acres in farmland and in forest, an average number of skunks in the woods, jackasses on the farms, rats in the cities, and electric carving knives in the homes. Each average is a most important datum—for those who love averages.

That there can be any alternative to this procedure is a strange and difficult idea for some people, previously unthought of, maybe even unthinkable.

Ordination

The methodology of *ordination* has its roots in the compulsive emotional behavior of viewing the world as composed of entities *each quite variable* within itself and definitely different from all others. (The careful reader will realize that internal variability, and external variability with regard to similar units, are not necessarily correlated; but I do not wish to make this discussion unnecessarily complicated. A unit may be part of a whole, and that whole but a part of a large whole, so the problem is specific only for any one Level of Integration.) The philosophy of ordination does not allow for groupings and orderly classifications. To the contrary, since no two samples are exactly alike, the worker can arrange his data on graphs and charts according to any, or several of the kinds of, data he chooses to take, such as length of ear-lobes, D.B.H. (the forester's term for *d*iameter *b*reast *h*igh), weight, or the number of offspring sired. Actually, by subjective and intelligent choice, the data are usually placed upon some reasonable base. For vegetation types, one uses soil moisture, precipitation, temperature, elevation; and then notes the correlations. By the act of ordination, we "prove" that discrete and separate units do not exist in nature. The result is inherent in the method. The method is inherent in an emotional compulsion.

The merging phenomena of the world about us are composed of relative continuities between relative discontinuities, like the changing cloud patterns of a summer sky. The method of averaging focusses upon the types and kinds, upon the continuities. The method of ordination focuses upon the infinite intergradations, upon the discontinuities.

The argument—of absolute continuities vs. absolute discontinuities —like many extremist arguments, has been an interesting spectator-sport. One takes sides. One cannot remain neutral. The combatants, in their zeal, have developed forceful and effective techniques. Like nations in wartime, there are technological advances that never would have been made in peacetime—but are extremely valuable after peace is declared, if ever declared. Actually, however, the argument is a foolish one, like a child's argument between black and white, when all the world is shades of gray. Both black and white are still useful as points of reference, even if they rarely, if ever, occur.

In closing this section, we should emphasize that neither averaging nor ordination can be made foolproof techniques, to be applied without the intelligence and judgment of the researcher. A man with one foot in the gutter is aware of a discontinuity in his life, especially if the other foot is on a banana peel. Yet the same man, when his marriage is breaking up, for a long time will not know which side of the fence he is on. Indeed, the idea of "fence" is inappropriate. It is even more difficult to analyze continuities and discontinuities in the world about us. Think of our frozen monkey again, that we want to study and understand, in terms of the scientific data we obtain (not by what we already know). We could take each slice, run it thru a blender or homogenizer, and then analyze each for a wide variety of materials, blood, nerve tissues, bone particles, proteins, fats, oxygen, carbon dioxide. And then (assuming we knew nothing of a whole live monkey), we could ordinate all these data using charts, graphs and refined mathematical procedures. The published papers would be truly impressive. I suspect, however, we would be missing the monkey himself. I doubt if all this prestigious scientific research would provide any greater understanding of the total monkey than the blind men obtained of the total elephant, as described in the Buddhist scripture *Udana*. An intelligent scientist will continue to speculate, not on what he knows, but on what he does not know. I know no sadder tale than what was told me by an eminent biologist, himself the sire of an ordination concept he called the "continuum" (that is, without discontinuities) as it occurs in vegetation. We were alone, in a quiet corner, during an international conference on the subject of "Man's Role in Changing the Face of the Earth." We were discussing his continuum. There was a pause. Suddenly, "You know, Frank. *I* do not believe in the continuum." He hesitated, while I grabbed my pencil and jokingly asked if I could quote him on this amazing admission. With infinite sadness, he continued, "Only my students do."

10. Methodology: Biometry and Variation

In section 8 above, we discussed the methodologic idea of sampling, of the taking of a part as representative of the whole. Then we obtain our sample data and assume—or hope—that it is representative of our hypothetical whole. We have no further interest or concern for the *un*sampled parts of that whole. To keep worrying about them would be a denial of faith in the sampling procedure that we did adopt; and such worry is not only a matter of not playing the game according to the rules, it is downright neurotic.

In section 9, we showed two techniques for using those data. The first was simple "averaging," the Goulash Approach in which homogeneity of a sampled natural phenomenon *is* obtained by stirring everything up in the same kettle. By definition, a mess stirred up using different sample-data is *another* kettle of fishy stew. Thus we have units that are *dis*continuous in nature, that can be arranged and classified to our heart's content, like foods on the shelves of a supermarket. The second technique was "ordination," the Angry-skunk-in-a-corner Approach, in which heterogeneity of natural phenomena *is* maintained by keeping each and every variation and component separate from each other. Whereas the Goulash Approach is fine for describing separate bowls of stew, ordination has its value where an infinite number of gradations deserve to be preserved. Where these gradations are arbitrarily and subjectively correlated with known gradients in nature, like altitude, summit-to-river topography, latitude and temperature, the correlations can and do become significant. But without some such guiding principle, without some application of human intelligence, ordination can be a fool's paradise. Our example of the frozen sliced monkey indicates that unless we already know it is a head-to-tail-arranged monkey, ordination of the slices would yield some very esoteric far-out biologic data, as far removed from the natural phenomenon as a non-Euclidian geometry applied to buying land.

In this section 10, we come to a third way of treating the data obtained from samples, although historically it was an elaboration and amplification of the Goulash Approach. This is the method of *biometric analysis*, also referred to as biometry, or as statistical methods. It is an eminently rational and practical application of mathematical principles that was first developed in the 1920's and 1930's. Biometric thought is grounded upon acceptance of the axiom that all biological phenomena exhibit *variation*, even as do our fingerprints. The description measurement and analysis of this variation is a matter of basic and practical scientific importance. As is true throughout science, we are confronted by subjective decisions at the start. At least two are needed in the simplest biometric analysis: the material samples, the "whole" (e.g. children of a certain age) ; and a measurable attribute of them, the "part" (e.g. their weights) .

A biologic variation, when graphed, exhibits the familiar "bell-shaped curve." By this term we mean that the bulk of the individuals will center around some average or mean. Individuals showing more and more extreme characteristics will flare out to form the tails of the graph. To discover the nature of this curve, for any one attribute of any one population, is itself scientific information of great interest.

Biometry "takes off" from this basic idea. Descriptions of the curve itself, whether it is tall and slender, or broad and squat, whether the tails are very long or not, can be described in terms of mathematical concepts.

The relations of one attribute to another can be compared, contrasted, correlated, or the attributes can be otherwise related to each other.

A single subsequent sample can then be evaluated as to whether it is an adequate sample or not, and how it might fit into another unknown total population.

Several samples can be compared and evaluated, with estimates as to the probability that they are, or are not, parts of a single whole.

Altho biometric methods cannot give us the "certainty" that some scientists crave and demand, biometry acknowledges the *variation* that does occur in scientific phenomena, analyzes it, and describes it in a variety of precise quantitative ways. Having done this, we are in a position to evaluate and place fragmentary phenomena which we know are parts of some variable but unknown whole or wholes, and which wholes can then be imaginatively and logically constructed. The intellectual approach is a distinct advance in our understanding of the world about us. Nevertheless we must not forget the array of underlying assumptions in the use of biometric methods, which *starts* with the idea that samples under scrutiny are from a hypothetical infinitely large population, and *ends* with the idea that we are only testing a hypothesis rather than initiating a discovery.

11. Methodology: Mathematical Formulae: Formulation of Laws.

There is no logical way to the discovery of these laws. There is only the way of intuition.

—ALBERT EINSTEIN (1879-1955)

To many natural scientists, certainly to the uninitiated layman, one of the most mystifying of scientific treatises is a textbook thoroly sprinkled with very impressive-looking mathematical formulae. Some will be reminded of the physics text books from which they may have memorized, but never really understood. Others, gazing on the strange and imaginative symbols, may wonder whether the roots of these ideas lie not in medieval astrology. Letters, numbers, symbols in complex relationships, some of the numbers of outlandishly odd fractions or decimals, all seemingly presented with the Mosaic finality of "natural law"—as indeed they are (so intended)!

Such mathematical formulation *is* a sound and logical scientific procedure, with extremely interesting psychologic undertones. Without necessarily "understanding" any one formula, it is in my opinion not too difficult to grasp the nature of the intellectual ratiocinations that are back of this seeming prestidigitation.

Notice that every such formula has an equality sign (=) somewhere in it. This sign can be read as "is," or "is equal to." In other words, a formula is a "sentence" involving the verb *to be*. For example, "two times two *is* four," which can also be written as "2 x 2 = 4."

We are now ready to accept the fact that every formula is *a relationship*, actually an *equal* relationship, an "identity," between what lies to the left of the =, and what lies to its right. Scientists are interested in relationships. The more it rains, the wetter the soil. The more it rains, the higher the level of the stream. The more it rains, the more quickly the garden plants grow. The more it rains, the more umbrellas are sold. Many of these phenomena can be expressed numerically.

If you have followed my line of reasoning, you will disagree, and say, "Oh *no!* These two sides are *not* equal. There may be a proportion, but not an identity. For every inch of rainfall, there may be only one-half an inch of plant growth." This problem, of course, is easily solved by such arithmetical devices as dividing or multiplying either side, or parts of a side, of the equation by some number. Thus our formula becomes more complex.

Still, you should be unsatisfied. You will maintain that there are often no simple fractional or decimal relationships between the Left, and the Right, there is some totally odd and irrational relationship, even if we can say that as the Left increases, the Right increases also. This is where the idea of k, or a "konstant," enters. (Spell it *constant* if you wish, but scientists think of it as *k*). They work up the data for a lot of Left situations, and then for the corresponding Right situations. And then they simply find out that in order *to* put an equality sign into this jumble, they *have* to find the number that *does* make the two sides equal. In short, the K number is created to *make* the two sides equal. Presto, we have a formula! A shorthand way of using figures and numbers that is far more compact and precise than a long complicated sentence of words, and that delights the mind of the mathematically inclined. Beyond arithmetic, algebra geometry and calculus can now be employed to create ever more ingenious expressions of natural relationships.

Let us return to the line of thought of preceding sections. When a suitable amount of quantitative data is obtained from the observation, experimentation, sampling procedures, and data manipulation previously discussed, it is always a problem to know what to do with it all, and to communicate this knowledge concisely and meaningfully. The formula serves this purpose. Once presented, it may be looked upon as a working hypothesis. It is tested in new situations. If it appears to hold, the formula becomes a natural "law." Essentially, the formula is nothing more than a description, in mathematical language, of certain situations which have occurred empirically with a certain uniformity, and can be counted on to appear in comparable situations (the subjective again!) with comparable uniformity. If they are found not to occur, *we change the law.* The law changes with the known events; it is simply a shorthand statement *of* known events.

One suspects that if it is not a strong emotional compulsion that leads some scientists to emphasize this methodology, it is at least the expression of a recognizable personality type. For example, one biologist who has more formulae in his textbook than any such author I at present recollect, discusses seven other methodologic approaches, for all of which he has adversely critical comments. Then he introduces formulation as *the* "rational method," implying that all others, if not irrational, are at least less rational. Formulation techniques he says are much used in engineering, are formulated on the basis of physical laws, and are recommended whenever there is sufficient physical basis for such formulation of functional relations. As with all methods, emotional enthusiasm may lead one to premature decisions based upon inadequate data. There is a special danger with formulae however. Once in the guise of a formula—like a devil daring to masque as a divine—few may recognize the deception, fewer dare to question it. The data fit the law? No, the law fits the data.

> *My object all sublime I shall achieve in time—*
> *To let the punishment fit the crime—*
> *The punishment fit the crime,*
> *And make each prisoner pent Unwillingly represent*
> *A source of innocent merriment!*
> *Of innocent merriment!*
> —WILLIAM S. GILBERT (and ARTHUR S. SULLIVAN)
> *The Mikado. 1885*

12. Methodology: Systems Analysis

(Critical metaphysical problems hide in the fact that the cybernetic system does not tell us what the goal should *be.) Once the goal is decided, the cybernetic system seeks it. As Louis Couffignal (1958) said some years ago, "Cybernetics is the art of ensuring the efficacy of an action."*
—HAROLD G. CASSIDY
in BioScience 17(12). December 1967.

For a considerable time I was playing a sort of experiment with some of my professional colleagues. Every time they would mention "systems analysis," I would show interest and ask questions. The tone of voice they all use when they first speak of this is a significant part of the picture. It is a direct analogue of what a minister uses as he repeats the ritual form of some holy sacrament. The subsequent conversation becomes sprinkled with such phrases as "general system theory," "simulation," "mathematical modeling," "cybernetics," "input and output," "parameters," "feedback," "rigorous quantitative approach," "elegant logic," and of course the indispensable presence of the digital "computer," listening to the pseudo-algebraic languages of Fortran and Algol, which serves as the altar for the practices of what bears all the earmarks of a new religious movement, replete with its evangelists and exuberant converts. Now I had often been asked myself "What is systems analysis?", and my fumbled answers gave me a distinct inferiority feeling. With the sincere intent of upgrading my intellectual image, I began asking those whom I considered to have the sophistication properly to answer. Quite honestly, it seemed to me that their answers always gave rise to embarrassment, for me, not for them. Their replies, when translated to simple words of one syllable, sounded like old ideas, not new. More than once, I ended up by apologizing for asking such simple questions (that received such simple answers). When a priest is explaining some higher point of dogma, it is a stupid if not discourteous layman who breaks in with "What do you mean by the Trinity?" I did once, however, find a scientist who, after the first unctuous declaration on the subject, admitted he did not really know, but there was a whole book on the subject which had recently been published. He had not read it yet, but I should do so.

Not too long thereafter, I found myself in the science library of a campus well-covered, at least figuratively, with ivy. The recommended book was open before me, to the first part, the first chapter, the first section, the first paragraph. I had the anticipation of one for whom the great mystery of mysteries was about to be unveiled. Then I began to laugh, I thought silently. When I looked up, it was a bit too late. for everyone within the room was staring at me, looking more as though I were guilty of maniacal chortling than simple laughter. (My embarrassment was worsened when I saw my legs draped over the corner of the table. (I prefer to do many important things in life prone, not on seat or feet; but I do try to keep my pedal extremities under the table when eating or reading, in public.) I do very sincerely recommend the most careful reading of this page, for all those who approach the subject of Systems Analysis with worshipful veneration. Unquestionably, the methodological techniques of Systems Analysis do enormously enhance our ability to describe complex natural systems in nature. Our previous mathematical techniques involved but two or three factors operating at some particular point in time and space. Here we can operate with a greater number of factors now called "variables," (all of seven, in the description I read), and with changing situations both in time and space. This is a toddler's step forward in the right direction. My humor arose from the fact that I immediately thought of a considerable number of highly important, practically important, parameters that were not of the Chosen Seven. When our science advances to considering seven hundred admittedly subjective variables, I may give the System my obeisance.

There is nothing new in the idea of a *"system."* We have already discussed the concept in the section on *holism* (which is not spelled "wholism"). It is little more than a natural phenomenon sufficiently integrated in nature to be considered worthy of study by itself, like the chemical reactions of a solution in a beaker, a machine in operation, an amoeba, a naked ape, a ghetto, a lake, an island, an earth. It has been in the scientific consciousness for all of this century, discovered and rediscovered by various people, at one time with all the pompousness of a self-proclaimed successor to Aristotle himself.

There is nothing new in the idea of "mathematical models." It is the very same idea we have just observed in the discussion on mathematical formulae, but elaborated and made sufficiently complicated so that far fewer natural scientists really understand them, and thus proportionally the more people worship—what they cannot understand. As with simple formulae, they are a highly abstract and simplified expression of a system based on the limited data we do know, and called a "model" (which has the subliminal attraction of both pure logic, and cheesecake), instead of a "formula" (which sounds like baby food). When the known facts of nature change, we change the model.

There is nothing new in the idea of a "cybernetic system." It is simply a dressed-up outgrowth of the old ecologic idea that every "subject matter" or holistic unit has an "environment" around it. That environment feeds something into it (like the ocean tossing a dead whale onto the shore of a coral atoll), and something going out of it, like a latrine built on a pier, or like a typhoon washing the same island clean of all animal life). Think of your own in-come and out-go, financially or nutrition-wise, and you will get the idea. The cybernetic approach recognizes that mathematically we can set up controls for the in-come, and can have undesirable changes in the out-go recognized by a "detector," thence operating thru a "governor" to alter the "in-put," so as to keep the system in a "steady state" which scientists prefer to call homeostasis (since no one else knows what that word means). It interested me that diagrams presenting these ideas appeared in 1967 in one of our leading science journals, presumably as a "contribution" in newer higher thought for the edification of the Scientific Estate itself. I would suggest they be incorporated in General Science books for the primary grades, as indeed they may already be, for these thoughts are old, simple, fundamental, and should be part of the educational heritage of every citizen.

 "When the known facts of nature change, we change the model."

There is nothing new in the idea of "simulation." We simulate a phenomenon whenever we do not for any reason wish to study nature directly. Thus conditions that are quite out of reach physically (those at the center of the earth), or intellectually (the complexities of urbanity) can be kindergartenized to the point where they are amenable to the paraphernalia of our offices and our laboratories.

There IS something new in that mathematical and mechanical t e c h n o l o g y which is being applied to implement these basic ideas. The more complex mathematical formulae allow us to bring more facets (variables) of the System into a single picture. (But we are still missing many many more facets). The technologic genius of the computer allows us to manipulate these data with a rapidity and a complexity that would otherwise demand armies of men, and years of time. The result is a superbly powerful technique. As said by Bill Surface, in *What computers cannot do* (Saturday Review 51 (28): 57-58, 66. July 13, 1968), "One new computer can make more computations in a single minute than a human mathematician could do by hand in 4,000 years." In addition to all these virtues, our old quaint and queasy quest for *prediction* can blossom out anew, with a beloved fury that makes the Roman haruspex, inspecting the entrails of a sacrificial victim, seem Neolithic by comparison. For example, with the adventurous growth of science and scientists during the 1950's and early 1960's, an irrefutable prediction could have been made for the year in which every man, woman and child in the United States would have been a science researcher.

> *". . . the computer has . . . encouraged*
> *. . . the belief that the things*
> *that count are those that can be counted."*

We see, therefore, that a technology has been a d d e d to some basic and fundamental concepts. As is true of all such situations in human society, there is the danger of the abstract concepts withering, and only the dry technology being left. The advent of the computer has often encouraged the trivialization of scholarship and the belief that the things that count are those that can be counted. Already Systems Analysis is beginning to dominate economics, at both corporate and government levels. The new nations believe it to be the solution to their under-development. Representatives of our nation, like the Peace Corps, find it is something they must "do," both by the request of the foreign nation, and of their own bureaucratic superiors.

For those, however, who can penetrate the facade of this prestigious technology, they will find, not science, but only a methodology of science. Once they remove the ecclesiastical garb, once they silence the droned intonations of the priestly caste, they will find some very frail and human judgments, weighed in the light of a most incomplete and inadequate understanding of the System which is man-and-his-total-environment, the o n l y System upon the successful homeostasis of which, depends the continuance of our civilization, of man as a species, indeed of the Earth itself.

An established method is the E N D, not the beginning, of creative scientific progress.

THE LITERATURE OF SCIENCE

To identify the age at which man first began to write is a semantic quibble. *What* is "writing?" Many animals leave scents and signs to notify others of their kind that they have been around. An analogous behavior pattern emerged in World War II with the sign "Kilroy was here." Individuality of the animal comes out when initials are written or carved, an innate pattern that has left school desks, fence bars, cave walls, and the strangest surfaces covered with these compulsions to establish individuality amongst an otherwise meaningless mass of organisms. The trick of a Scout not only to leave such a symbol on a trail, but to indicate as by an arrow which way he went (or even to mislead by indicating the wrong way) is also a form of "writing" that goes back to our mammalian forebears. Both predator and prey, both friend and foe among animals, quickly understand which animal went which way when they see a sign, and I suspect with far more certainty than the Scout who thinks he stands at the high point in the evolutionary "Descent of Man," if I may use Darwin's directional term.

When man first began to communicate complicated ideas with pictographs, ideograms, or phonetic alphabets, civilization entered upon a new and fruitful era. For the first time, accumulated knowledge and culture were no longer dependent upon word-of-mouth transference. In verbal communication—as continued with the bards and minstrels —knowledge was subject to gradual change with repeated tellings; and one break in the civilization-chain left the knowledge totally lost. In written communication, however, knowledge was not subject to gradual change; it need not be lost during dark ages; it could be learned at the leisure of the individual. Most important, one individual could "talk" to m a n y, even tho talker and hearer were widely separated in space and time. Despite the superb intellectual accomplishment of w r i t i n g, I doubt whether its nature has become imbedded in our genetic constitution. How many people, even scientists, prefer to talk at cocktail parties and conferences, than to read? How many homes give the place of honor to the talking television set, rather than the silent bookshelf?

The "literature" of science is its accumulated writings, its wisdom and non-wisdom. It is vast and voluminous, and totally beyond the ability of one man to master in a lifetime. Even the contemporary flood from the printing presses is too much for one scientist to keep abreast of. Thus the need for popularizations, for "abstracting journals" (which give full bibliographic references, and short condensations), and annual surveys of special fields. So great is the flood that one wisp of hay is the more easily "lost." Thus, we are drafting computers for what is solemnly called "information retrieval," to act like the canine retriever who can go snipe-hunting and really find one snipe in a large marsh. In all the joys of automated computer retrieving, we are entirely likely to overlook the very human limitations of the technician who programs the original publication into the computer. A book on "intestinal gardening" for example, might be buttoned into the organic gardening and landscape fields, whereas the subject concerns the control of the bacterial flora in connection with human laxity.

Most scientists pay tremendous deference and lip worship to "the literature," particularly in their dealings with students. Nevertheless, we cannot completely overcome our anthropoidal ancestry, and its emphasis on word-of-mouth communication. The never-ending succession of symposia, conventions, conferences, congresses and conversational confabs testify. There are diarrheic outpourings of verbosity, which the participants rationalize as "important" by perpetuating the more substantial material in voluminous printed "Proceedings." This solid state, more often than not, either repeats what the authors have already published once or many times, or the nugget of which is fully contained in the abstract of it. If the matter of acquiring new knowledge were the main purpose of these meetings, one would stay at home, read or scan all the Proceedings, and glean from them what one wished in a very small fraction of the time and expense involved in attending the meetings. But we are social animals, and we rationalize in accordance. By nature, reading is difficult and unpleasant; talking is the easier. I have had only one eminent colleague honest enough to tell me "Whyinhell should I read all the stuff you have published when you and I can spend a whole day together talking?" (*He* was willing to spend such a day. I publish to *avoid* such a day.) Yet I admit that I am repeatedly tricked into spending such days, the less so the older I get, as I realize that time is running out, time to write, not to talk to people who won't read and who will soon die anyway. Sure, there are advantages to talking, which seldom is con-versation in its strict etymologic sense, particularly if one party has never learned how to talk—or to read—much less to listen. The man who reads, providing he is intelligent, will always learn the more.

The literature of science may conveniently be divided into three major groups:

> *The Professional Research Literature*
> *The Professional Education Literature*
> *The Popularized Science-for-the-Layman Liturature.*

The Professional Research Literature

This is the literature to which most scientists refer when they talk of "keeping up" with it. And when they so talk, you can be quite sure they are n o t keeping up with it. There are essentially two types of publications here: the research article journals, and the news journals.

The *research article* journals are the media by which a scientist presents to the world his "original" contributions to new scientific knowledge. These are the journals which are at the very crest of the wave of advancing scientific knowledge. I am sure it does not detract from this very holy role for me to say that in diction, phraseology, idiom, style, format, and development of thought, this literature is as ritualized, as ironclad in its rules, as predictable and stereotyped as any of the liturgies of the ancient religions. As is true of the social behavior pattern of any group that seeks superiority and distinction by becoming unintelligible to all other groups, scientese is peppered with impressive patter, that parallels for example the jargon of physicians. They pride themselves on tightness and conciseness, on using the fewest number of words. Graduate students usually find that when they write their theses, the first draft comes back with the comment "cut it in half." This diminution may happen more than once. Then the editor of the journal will make other lengthy deletions. Eventually the published article may look like an abstract of the original. Nonetheless, the scientist will sprinkle his writings with "technical" terms, which are merely translations of common words into others of unusual Greek and Latin etymology. It is not apocryphal that one ecologist could not call a spade a spade; he wished to call it a geotome (that which *cuts* off a piece of *earth*). This literature is dogmatically shorn of everything but facts, but data, but truth and reality. It is abhorrent to allow anything to slip by that could be damned as emotional, a value judgment, or a personal opinion (even tho many of these technical terms are outrageously ecstatic when translated into Anglo-Saxon, to wit, all the -phily terms; entomophilous plants are "insect-loving" plants, actually those which are pollinated by the agency of insects; but surely a broody female flower could be excused if it showed affection for a pimping pollinating bug).

The greatest of all sins in scientific writing is what is branded as "anthropomorphism," the ascribing of human characteristics, of purposes and feelings, to other organisms. All other organisms are unintelligent un-feeling, un-conscious products of evolution that are blindly reacting to their inheritance and their environment. I have long predicted that a coming scientific generation will adopt a valid "zoomorphism," the ascription of animal intelligence and feelings to the otherwise unintelligent behavior of human beings blindly reacting to their inheritance and their environment. Perhaps it is here already.

Second only to the sin of anthropomorphism, and in a sense a part of it, is the sin of teleology; the lady without whom no biologist can live, yet he is ashamed to show himself in public with her. Teleology is an interpretation of natural phenomena in terms of purpose, design, intent and foresight. Woe betide the biologist who is so unsophisticated and unindoctrinated as to write that an animal does something "so that" something else happens, or "in order to" bring about some other event. A bird does not build a nest "in order to" have a place for eggs to be laid and nestlings to be raised. The bird is a mere puppet of behavior patterns mechanistically determined by his inheritance and his environment. But have you ever watched hungry gray squirrels in winter trying to reach the bounty of a bird-feeder, even tho it is suspended from a l o n g wire attached to two distant supports, and even tho the wire by itself is a veritable obstacle-course of roadblocks, rolling cans, and long-quilled deterrents? One need only watch the little beast below, brightly eyeing each and every element of the system including the nearby limbs of trees and eaves of roofs, before scurrying up to take a flying leap from the nearest perch, with a precision that would do credit to computerized pinpoint bombing. Here is purpose design and intent, with an observable see-the-wheels-go-round appearance in their sage faces that is too often missing from the visages I see so often from the lecture rostrum. Notwithstanding this taboo on teleology, purpose design intent and foresight are the veritable sacraments of man's faith in man. Why not? Man is the pinnacle of creation, as man's science has proved. Man alone has been granted intelligence. (Thank God no other animal has it.)

Another characteristic of the scientific research article is that regardless of how the ideas were developed, or what roles insight, intuition and hunches played, or of how the data eventually supported the insight, it must, nevertheless, be written in strict conformity with the "logical" procedure of the scientific methods described above: the problem, the methods, the data, discussion, conclusions. The sequence of parts is as rigidly determined as the sequence of the parts of a liturgy. Editors insist on it; the authors fool themselves, though occasionally an honest scientist will openly question the integrity and the ethics. No use to do so; the dogma decides differently.

It is important to realize that research journals are open only to research. Entrance within the pearly bindings is based on one criterion: Is it "original research?" Is it a "new" contribution to human knowledge? There is no more insulting condemnation from an editor than for an author to be reminded of this, if he is fool enough to submit a paper of any other category, no matter how important it may be to the field of science.

The one possible exception to this iron-clad rule is in the matter of "reviews." Most editors look upon this part of the journal as ideally one for reviews of "original research" published elsewhere, written in a style that results in little more than a condensation and abstract. Nevertheless, the horny sin of devilish judgment and opinion always raises its attractive head, and seeks entry into the virtuous unfeeling body of the research journal. Review Editors are only human; they sometimes slip, to the enjoyment of their readers, only to feel such guilt that they quickly return to a cold collection of facts.

The *news journals* are by no means restricted to "original research" tho they may and often publish a considerable amount. They do publish a number of general, and timely controversial articles. Altho the language is still scientese, there is considerably more freedom of subject matter, especially as to review articles, and coverage of current events. Since editors rather than research scientists are now in positions of administrative power, and since editors are frequently frustrated writers, often entering the field sidewise not from science but from the humanities, mistakes can be made that are ludicrous, embarrassing, and even highly destructive to the scientific community.

The Professional Education Literature

In accord with a segregation we have previously noted, this scientific literature breaks down into (1) the education-dominated literature of the schools, and (2) the science-dominated literature of the colleges and universities.

The education-dominated literature of the schools mainly revolves around (a) the journals of science-teacher organizations. They have much in them that is eminently sound. The emphasis is on the teaching methods, and beyond the immediate scope of our present interest. This literature also revolves around (b) the school texts, again more education-dominated than science-dominated. Changes are constantly occurring in this literature. One of the most revolutionary of these changes was in biology, in the 1960's, the program known as the Biological Sciences Curriculum Study. This hugely expensive project, the result of cooperation between professional scientists and high school teachers, resulted in three new volumes, volumes that were vast improvements in the matter of textbook form and style, even tho the actual quantity and level of scientific information probably exceeded the capacities of the age-groups involved, as well as of the teachers themselves.

Secondly, the science dominated literature of the colleges and universities mainly revolves around what is called the "textbook." The textbook author is the Castor for which the "lecturer" is the Pollux of these stars of the academic firmament. In more ways than one, the textbook is rightfully called the Bible of a course. It is written with sufficient ambiguity and obscurity, with enough unintentional errors, with enough incompleteness (in the lecturer's opinion), with such atrocious pedagogical design (as in not having a unit of text and pertinent illustrations and graphs on the same page or on facing pages) that a minister of the faith is needed, not only to choose a text for each sermon and expound upon it, but to translate it from scientese into the language of the laity.

The textbook author is a very special personality-type. Fortunately, there are very few of them. They have the self-confidence and know-it-allness of a self-proclaimed prophet. The degree of this attainment is very often correlated with a type of stability within the author's mind. More than one accepted textbook can be recognized from the lecture notes that another scientist obtained from the same course as that in which the author was introduced to his subject. Textbook authors do keep abreast of the times, in their manner. But this breast is very often recognizable as an external padded support, a false front designed more to give a modish form to what would otherwise be a senile sag.

The "textbook publisher" is an integral part of the conservative forces which perpetuate this aspect of the durable and inflexible academic world. Publishers, like all other corporations, have but one god, Mammon, despite what lip-service they may pay to other parts of the pantheon. Understandably any manuscript proposed for publication must come up to certain standards, or it stands no more chance than did Jesus among his Jews. The manuscript must be approved by anonymous reviewers among the author's peers (who are probably competing textbook authors). The proposed book must fit into existing courses, ones already being taught. If it demanded a new and different course, there would, simply, be no sales. Few professors, even if innovative when they start teaching and most are not, can change in the mid-stream of life to a different communicational chalice. The result: "new" textbooks that are only restamps of the old tradition, with a sprinkling of new "facts" that are already out of date before the book is off the press. When these obsolete "facts" become too obvious—or even without—publishers put out new "editions," often reprintings with very minor alterations. This gimmick makes obsolete and worthless all past editions, increases sales for a temporary spurt, and does bring up serious problems of social ethics—problems which remain as long as the corporation is king.

The Popularized Science-for-the-Layman Literature.

Most scientists are not interested in communicating to the layman. Indeed, the layman does not "exist" for most scientists, so closeted is he in his ivory tower. Yes, he sees the masses, on a summer beach, in a city street, but to communicate to *them* is entirely alien to his nature. He communicates only to college students, on the assumption that they too will be "called" to this most noble of all lives. Indeed, to be interested in communicating to the hoi-polloi may well endanger the standing of a scientist among his peers. If such i s his interest, he is a mere "teacher," a lesser kind of mortal who may sweep the temple and hand out bread to the hungry, but may not officiate in the services.

For these reasons, the finest popularized science is often produced by journalists, by writers, for whom language can be an instrument as fine as a Stradivarius, to be played upon with an esthetic sensitivity as subtle and as perfect as that of any master artist. You can imagine how the Old Guard, steeped in the rituals of their own archaic dialect, react to the freedom and appeal and beauty, (even tho there is also scientific correctness) of these science-writers for the people. One of the most fascinating types of literature written by scientists is in their reviews of such books, reviews written with cutting and condescending snideness, accusing the author of all manner of excess verbiage, of emotionalism, of inexactness and indefiniteness.

Of two general types of this popularized science literature, one includes the heavily promoted, gaudily illustrated, relatively weak-texted, seldom read, coffee-table-ornamenting, series of the large publishing houses. There is no doubt about it that these are g o o d books. For those who will read them, one will find that they are infinitely superior in design and format to the stereotyped classroom textbook. They are meant to be understandable with*out* an exegetic professor. Illustrations and explanatory text are so linked that the eye may cross from one to the other without the interruption of page-flipping. And the very quantity sold allows for art work and photographic reproduction, far beyond the means of a separate one-author volume. Unfortunately, authors must generally be "hired" for this sort of writing, with the inevitable tendency to decline into simple hack work.

The second type of popularized scientific literature—as with all creative artistry in civilization—is the result of individual genius, even if hack aides fill in inconsequential details. It must be said, however, that so exacting are the scientific demands, so refined the artistic demands, that such individuals stand out less in the contemporary scene than in all of history. In retrospect, it will be found that most of them are "bi-aptitude" people, those who might have been equally successful in either field. They may have chosen science first, were steeped in the rigors of its factual and logical training, rebelled in conflict with the unbeautiful functionalism of an essentially beautiful human experience ,and finally burst the restraining pinions and soared forth, to give visions that will everlastingly ornament our understanding of the world around us. So, was Rachel Carson.

Human history more and more becomes a race between education and catastrophe.

—H. G. WELLS (1866-1946)

SCIENCE AS A SOCIAL PHENOMENON

The most scathing comment on our pervasive intellectual system, one com-
pounded of hypocrisy, self-righteousness, self-delusion and cliché, is that an
idea which could have been inspired by little more than common sense
and common observation must be labeled as new, as revolutionary, even as
Too Hot to Handle.

—ROBERT ARDREY
(in a book review, Life magazine June 20, 1969)

In the preceding sections of this book we have discussed *science* as
a phenomenon of its own Ivory Tower, very much as we might consider the nature of a monastic order, its principles and ideals, and the
activities of its followers, its proselytes and its apostates.

Our interests will now broaden, as we consider several facets of
science as itself a social phenomenon, and its interrelationships with
other social phenomena. Altho scientists study everything under the
sun and thru the universe, they rarely study themselves. Consequently,
our knowledge about science and scientists is very remarkable—for its
absence. Sociology has shown little interest.

1.   Fads and "Schools" in Science

Fads, hobbies and crazes advance thru human society like local
thundershowers on a summer day. Suddenly everyone is under an
umbrella. Just as suddenly, the umbrellas have disappeared. So it is
with song hits, with stars of the stage and screen, with hemlines and
necklines and problems of revealed knees and ankles and breasts; and
so it is—scientists being human—with the revealed truths in the science of the times. The general public is usually unaware of these
passing fancies. I doubt if scientists themselves are fully aware of the
degree to which they are motivated by such ephemeral social forces,
which certainly do no justice to the "logical decisions" and "intellectual freedom" which they believe they demonstrate. On the other
hand, there is certainly no disparagement in admitting that styles and
tides wax and wane thru the whole inorganic and organic worlds. It
is so even for the birds. There was a time when some bright innovative
little tit of a street bird of England discovered it could lift the cap of
a milk bottle placed in the early morning on the kitchen doorstep, and
quaff deeply. Other birds observed, and mimicked. This very practical
avio-scientific technologic discovery spread across all southern England. In an enterprising flight across the Channel, the methodology
advanced in ever enlargening waves thru the Low Countries. I am
not aware that human scientists have recorded the fall of this fad. I
assume it is related to responses from the environment, such as changes
in milk containers, or the earlier setting of alarm clocks in households.

The fads in scientific interests would first be evident in the proposals for financial grants from the philanthropic foundations and the government agencies. Such social units would be particularly sensitive to such expressions of mob psychology, since the suppliants would mold their requests in the manner that would most likely bring the dollars jingling out. I am not sure any such study has been made. I once made an effort in this direction, with a preliminary "Proposal to Study Proposals." Since the huge granting agency involved was short of funds that year, with requests totalling 1000% of available moneys, and my proposal was to cost only .01% of the total proposals received by them for a single month, since two friends of mine were in administrative capacities at the time, and were not only sympathetic to my idea but had already been asking for my evaluation on several proposals and knew that my objectives would still further assist them in their program, I felt reasonably sure that my idea would be seriously considered. The reaction came by telephone, with a preamble of flattery and praise, and then a rejection of such palpable transparency that I was more amused than annoyed. Amongst other suggestions that would make it appear that the door was not firmly closed, it was suggested that I obtain a co-worker in a totally different field. True, mine was obviously a bastard scientific study; but it was also very clear that I was in the same category as my proposed study, and that any formal proposal from me would have given the agency the perfect excuse to say that my objective did not belong at all in their agency, but in a totally different one. Clearly, they anticipated from me a study and a report that might have jangled the nerves of this multi-million dollar sugar-daddy, who was shelling out the shekels in support of his private nunneries.

Another type of social development that would allow for an understanding of the fads in scientific interest lies within the research journals themselves. Altho within the publishing flood of the 1960's, it is not considered fashionable to dig back in the literature more than 10 years (it is said to be cheaper to do the research over again, than to find out if it has been done before), or even 20 years, nevertheless if an undergraduate student were to blow the dust off the older volumes, he would be amazed to find how the times have changed. I strongly recommend such social studies of science, tho alas I fear it is only the historians who would make them and who would be interested in them.

*   *   *   *   *   *

A social phenomenon related to fads is that which centers around a single outstanding individual, and two or three generations of followers. Such a group is known as a "school." Sooner or later the group splinters, breaks up or dissolves, to reform into new schools and movements. All the world's great religions appear to have been born in this manner. And when we look into the minor sects, into the "heresies," into the fringe groups that are disparagingly called cults, we find the same sources and origins in the dynamic guidance and leadership and inspiration of a single man. Sometimes a second leader will emerge, building upon and extending the influence of the first, as with Paul, and Asoka. After the initial group of disciples come the phalanxes of followers, the blind followers, and the camp followers. Ritual becomes meaningless until reformation bears a new ministry. There are strik-

 ". . . the second and third generations . . . an empty shell, repeating the form but not the substance."

ing parallels in science, operating today, visible in the current research journals, and offering superb opportunities for the social critic. Alas, all too often the second and third generations adopt the mannerisms of their master without his inner grasp and understanding. They use the methods, the same words, but the intellectual breadth is gone; an empty shell, repeating the form but not the substance. And so these cycles come and go. After all, great men are rare. Others only follow the leader. Think what chaos and anarchy there would be if each acted as a leader, and there were none to follow even if blindly. We should be slow to criticize scientists for being human!

Ashley L. Schiff

FIRE AND WATER. SCIENTIFIC HERESY IN
THE FOREST SERVICE

Cambridge, Mass.: Harvard University Press. 225 pp. 1962

Ashley L. Schiff, an historian, has given us a remarkable documentation of the social problems of integrating into society certain scientific knowledge (some of it so simple and obvious it could be called "common sense"). His is a brilliantly damning case history, involving the U. S. Forest Service and their conflicts over the rational use of fire in forest management, and (to a less extent) the significance of deforestation in causing drouths and floods. His discussions are based on extensive interviews with living people, and research into correspondence files that are preserved in forestry libraries. The extensive quotations from letters are a goldmine of illuminating empiricism.

The author interprets his subject largely as a dispute between the administrative (applied forest management) and research branches of a federal bureaucracy, and one that might be solved by a more rational understanding between those branches. On the other hand, the message of this study deserves to be placed in a far larger context. Here are displayed all the emotions, unintelligence and prejudice of normal bigoted human beings. Here is good and evil, in the starkness of black and white, supported by a host of irrational compulsions. Fire is a short-term evil, and must be rooted out by all the techniques of a permissible inquisition, including insidious suppression, and propaganda worthy of P. T. Barnum. Reforestation is good, and the evils of drouth and flood must be due to deforestation. The few crusaders (and that they are looked upon as crusaders is the most damning part of this study in social psychology) may have common sense on their side, the experience of the land-owners themselves, and the facts of documented research. Nevertheless, they face an over-population of Smokey Bears as well as the unimaginative stodginess of research technicians who for various reasons will not ask the proper questions. What happened, and is happening in the U.S. Forest Service will continue to happen in land-management bureaucracies. May Ninth Level ecologists be forewarned, which is to be forearmed.

2. "Knowledge-Flow" of Science

We have already considered several important aspects of the flow of scientific knowledge. We have discussed it from linguistic and psychologic viewpoints. We have considered both popular and professional literature, and the growing, feed-back, and diminishing features of "schools" and fads. We will now review and amplify the tale, for some of the most critical social problems lies not only in the flow of sound scientific knowledge, but in the flow of *un*sound scientific knowledge.

1. **Knowledge-Flow to the Younger Generation.** Under the general term of e d u c a t i o n is embraced all that communication of scientific knowledge to those younger than ourselves. In the field of education, two extreme approaches are theoretically possible, and periodically each emerges and dominates the social scene.

The "Far Left" may be characterized as pedagogy in its purest form when content means little or nothing, but everything depends on the methods. (This is as if the sections on methodology, pages 35 to 82 were 99 percent of this "science," as indeed they are in the thought and behavior of many scientists). This situation once dominated American secondary education, rallying around the name of John Dewey in the second quarter of the 20th century.

The "Far Right" may be characterized as "research, not teaching," and is present in those grooves of academe where the publish-or-perish principle is paramount. By this principle, a professor's personal security—which is, after all, dominant above all his other quaint and queasy quests—depends upon his research activities. The retention of his job, advancement and tenure, are keyed to research, and thus to his international standing among his peers. Teaching ability counts as nought.

The teaching of students is essentially a one-way transaction, with a captive even if obstinate audience. It is often analogous to the filling of empty pitchers, with no competition or comparisons with other and related sources of knowledge. Who ever heard, for example, of two professors holding forth in adjacent rooms on the very same subject, and competing for students, if not by their eloquence, then by such fringe benefits as beer and pretzels, or wine women and song? Furthermore, where the pitchers are already filled, and where the motivation of the audience is grade-getting and degree-getting, the technique of regurgitation-on-examinations fills all the terms of the Educational Contract.

Those who look upon education as a form of animal behavior are well aware that age and educableness are inversely correlated. The pitchers are most empty in the first year. Consequently, there is a paradox of subtle humor that society puts least time attention and money on the most formative years. When the little brats first enter school they not only have already learned a lot from unwise parents and from the streets, but they are already beyond un-learning and re-learning. The greatest time attention and money of society goes to later educational efforts that are as futile and foolish as the worn gag about trying to teach an old dog new tricks.

Before leaving this subject of Education, a word is needed in mention of a technologic field which in the 1960's was pouring forth and inundating our school systems like some tremendous lava flow. I refer to "automated learning." In this sociologic development are the enormously expensive "teaching machines," outgrowths of the fertile brains of the engineers and the technologies, boosted by our economic and industrial systems, and eagerly grasped by pedagogic professionals who claim that fewer teachers can thus teach more students. (Perhaps just as well, for more and more of them cannot teach anyway). I do not deny that such machines have a definite and valuable role in the field of Education. For example, to learn to drive an automobile while sitting in a classroom model, subjected to simulated dangers and vicissitudes of actual driving, rather than learning on the highways and killing ourselves in the process (as long as society has no suitable control over the autonomous highway agencies and automobile industries), this mode of teaching is clearly desirable. Nonetheless, every time I see illustrations of fully automated classrooms, a cold televised image of the professor in front, each child isolated within his own cubicle, ears covered with headphones, eyes filled with mechanical gadgetry, his own classmates isolated from him, I cringe at the cruelty, and marvel at the inhumanity of man to man. The student-teacher relationship, an extension of the parent-child relationship, is one of the most warmly personal in all our lives. To be a teacher is to answer a Call, not to take a job for the sake of the high salary it offers. The present ideas on automated learning are technologically sound; they overlook the fact that we are human beings.

2. Knowledge-Flow amongst Scientists. "Success" for a scientist is generally measured by the degree to which he is "accepted" by that specialist-group he chooses as his own, and which thus become his "peers." There are many ways of climbing this social ladder of success:

One way is by publishing in the professional journals. There are hazards on this road, for not only are the journals poor in funds, with

a back-log of papers to publish. The editors may exert a degree of
personal dictatorship equal in its way to the greatest of our political
dictators. There is a system of "anonymous reviewers" of manuscripts,
whereby your worst professional adversary can damn your paper with
insulting comments, and the author has absolutely no redress to the
editor, much less to the reviewer. Then an editor can delete what he
does not like, including what he calls "controversial points" and such
canine teeth, or he may even add sections that the author would never
write himself. Or the editor may place the manuscript on a shelf, with
purposive pèrmanence thus killing it, or until that casual day when
he may have an empty space in his journal, by which time the manuscript may be totally out of date and require complete rewriting, if
indeed it can stand resurrection. All this time the poor professor is
languishing in publish-or-perish fears. The readers of the pages eventually printed are totally unaware of all this meddling, for the editorial
deletions and additions are not indicated. Indeed, the author himself
may know nothing until he reads in print the castrated or monkey-
nut-grafted version of his original manuscript. When medieval scribes
meddled with the authors of antiquity, at least those authors were not
around to defend themselves. But what future scientist, when he digs
in the literature of today will understand the perversion of this cooperation between author and editor?

By this statement, I do not wish to imply that authors are always
guiltless. Editors have real problems with authors who cannot write
their language, and whose research and ideas are truly inadequate.
Science grows slowly, and it is difficult if not impossible to separate the
crackpot from the man ahead of his times.

In addition to publishing in professional journals, scientists enjoy
a wide variety of social gatherings—tribal rituals of the profession—
under the designations of *meetings, conferences, symposia,* and *seminars.* These can be extremely important affairs, primarily for the inci-
dental personal contacts that may take place in halls, in sleeping rooms,
at meal-times, or late at night; or for the subsequently published
papers which thus reach a far larger audience than the handful at the
time, and for which publication the paper was basically intended anyway. The motivation for these festive gatherings is pure animal soci-
ability. As a learning device, they are relatively ineffective. Those who
can read, learn far more by reading thru the same amount of time.

"Meetings" can be huge affairs, sponsored by one or several societies,
with many programs running concomitantly. Papers are usually read,
literally read, sometimes with advance copies in the hands of the
audience itself. Since scientists are usually poor readers, this formality
might be dispensed with, except that it is ostensibly the raison d'etre

of the program. Discussion time is often reduced to zero, since programs generally run behind time. Attendance at these meetings includes many types of people. There are the administrators who come to administrate, and who jockey for that supreme honor of a lifetime, the presidency of their major scientific society, an honorary one-year position which may involve little more than giving a non-controversial address at the end of their term of office. Then there are the oldsters, who come to impress the youngsters; and the youngsters who come to impress the oldsters; and the jobless looking for a job. Those who are scientifically busy may not waste the time to come.

"Conferences" are designed around a special topic, and take place independently. Attendance is by invitation; specific subjects are often assigned, and the general caliber can or should be relatively high. Insofar as papers are verbally read, they may differ little from programs of the meetings mentioned above.

"Symposia" are special programs, often within the programs of large meetings. They are designed around critical subjects, on specific aspects of which supposedly eminent authorities are invited to speak, the eminence of the honor of which usually carries expense-paid trips for the participants. There is frequently a real need for resurveying and reintegrating knowledge about a specific subject. All too often however, symposia disintegrate into opportunities for various Prophets to reaffirm their sectarian messages, in the hopes of gaining new Disciples. They talk, but do not listen to others, much less learn. Like great ocean liners in heavy harbor fog at night, all lights on, all horns blasting, they succeed admirably in avoiding each other. If contact is accidentally made however, the result is disastrous to either or both parties.

"Seminars" are generally open impromptu discussions, around chosen topics, by small groups specially invited. These might be looked upon as organized and directed bull sessions. If a man of iron is at the helm, of tact and brains as well, seminars can be amazingly constructive for the parties concerned.

In general, I would say that scientists prefer to talk, not read.

3. Knowledge-Flow to Society. Knowledge-flow to a classroom is essentially the teaching of a captive and hopefully obedient audience. Knowledge-flow to one's colleagues is a jockeying for peck order within the local cackling flock. But knowledge-flow to the outside world is a vastly different phenomenon, for which the teacher and the scientist are often woefully unfit by temperament and sophistication.

There are two major motivations for scientists to seek to communi-

cate with society. The one is the personal profit motive, a motivation
which may remain deeply hidden beneath their scientific rationalizations, a motivation which after all is fundamental to the American
form of government (with its system of taxation upon income) and
one which frequently crops out in cases of medical quackery even if
in part sincere and well-intentioned.

The second motivation is an altruistic and selfless interest in the
good of society. Even tho society would greatly alter without this
minute minority, such people are so rare and so regularly misunder-
stood that we still frequently jail them, even tho we no longer crucify
them.

Three facets of the problem deserve special recognition: the source
of the knowledge, the route of the knowledge, and the recipient of
the knowledge.

The *source* of the knowledge (if there is a source) lies in the original research, to which subject we have already given so much attention, itself depending upon our sensations of the external world, and
our mental perceptions of them.

The *route* of the knowledge lies in several well-studied media of
mass communication in society: radio, television, cinema, the press—
and the schools. In the press, newspapers and newsstand magazines
are the chief elements. Along with these media, and using them, we
must consider the fields of advertising, and public relations. These
two social phenomena cut huge financial slices in industry financing,
and thus become formidable adversaries in techniques and procedures
for those scientists who not only have no such funds to employ, but
who may be logically silenced if their universities are supported in
part by funds from these same corporations. The "corporate profit" is
but an analog of the "private profit," and neither need involve the
long-term good of society.

> *"Audiences go where they
> learn what they want to learn."*

The *recipient* of the knowledge is that obstinate audience already
mentioned. Audiences go where they learn what they want to learn.

Here also the personal profit motive is a factor. An audience will not learn what is a threat and danger to its own security. Such an audience will only brand the communicator as biased, prejudiced, ignorant or even fraudulent. Recipient audiences are not only individuals and citizens, they are also "social units" of every rank and category, in government, industry, and civic groups, in medicine, religion, recreational and artistic fields.

It is when potential recipients of knowledge prove refractory and obstinate that the fine arts of manipulating human behavior are brought into play. It is here where the sciences of anthropology, psychology and social psychology—and the anthropologists, psychologists and social psychologists—develop their knowledge and abilities with horrendous efficiency. The persuasive tactics of the Inquisition were archaically crude. In lieu, emerge all the subtleties and success of a man with a maid, from con man thru salesman, advertiser, public relations official, to the tactics of the conglomerate corporation itself. Their recipient-victims receive that knowledge of which they choose to be the source.

Source, route, recipient, and the purposeful influence of each upon the other two, force us to reconsider our subject not as one of the "Flow of Knowledge to Society," but of the "Flow of Sound and UN-sound Knowledge thru Society." Here is a field of knowledge that awaits empirical description and interpretation by those ethnologists and anthropologists who have conventionally restricted their interests to savage tribes and uncivilized races, and who alone are capable of describing odd irrational and self-destructive behavior with total objectivity and without being themselves influenced by the moral codes and religious beliefs of the tribes they are studying. Figures 1 to 3 represent the results of initial studies in chart form. In each case, a wide variety of "social units" were studied by questionnaire and interview as to the knowledge that each embraced. The data were organized to reveal the role that each social unit played as "source," as "route," and as "recipient," both for "sound (blue line) knowledge" and for "unsound (red line) knowledge." At times, red line knowledge had to be divided into mutually exclusive extremist groups, then referred to as a "far left" and a "far right." Within this framework, the charts were prepared.

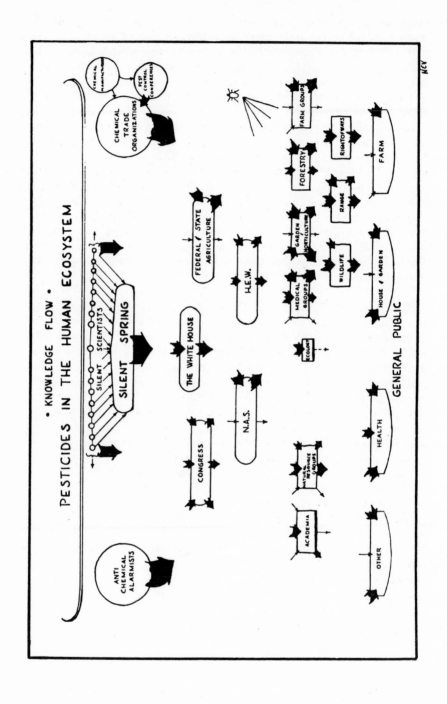

FIG. 1. *Flow of Knowledge Concerning Pesticides in the Human Ecosystem.* Chart based on a survey made in 1962–1963 of all available literature from the social units involved, including articles and notices in newspapers, magazines, television, and radio. The social units were of the following major categories: the "silent scientists" supplying much of the basic ecologic knowledge; the book "Silent Spring" by Rachel Carson; a group referred to as "anti-chemical alarmists" including the extremists of the left; the pro-pesticide industry of the right (including chemical manufacturers, pest control conferences, and chemical trade organizations); government (in which the principal actors were the White House, Congress, federal and state agriculture, the National Academy of Sciences, and the Dept. of Health Education and Welfare (including the Public Health Service, and the Food and Drug Administration). Then there was a large group of social units forming three separate clusters: (1) our colleges and campuses (excluding agriculture); (2) a group that might be called ecology, ecologists, and the Ecological Society of America; and (3) a complex of seven, involving medical groups, gardens and horticulture, forestry, farm groups (these four are tied to each other, and heavily influenced by entomology), wildlife, range, and rightofways. Lastly are the general public and the individual consumer, among whom we must segregate those of the farm, of house and garden, the ones concerned with health, and a catch-all for the remainder. Three types of knowledge must be recognized: (1) that of ecosystem ecology and general ecology, represented by *vertical* arrows; (2) that of the anti-chemical extremists, represented by *diagonal* arrows of the *left;* and (3) that of the pro-pesticide elements, represented by *diagonal* arrows of the *right.* The role of each social unit in the flow of pesticide knowledge is indicated by the size and kind of arrows entering and leaving that social unit, as estimated by the author as a result of his studies. (Actual connecting lines between units are not shown, because of their great numbers.) The chart is extremely revealing of the roles in our society of the three types of knowledge. The overwhelming abundance of heavy arrows of the "right" is everywhere evident, not only in government, but also in the general public. The extremists and alarmists of the "left" have made little impression on government, little impression on other social groups except for gardening and horticulture, but considerable impression on the general public. The role of "vertical" ecosystem ecology is the most instructive aspect of this chart, for it is in this direction that the fundamental scientific knowledge is moving. At the top of the chart we find the "silent scientists," who supplied their information for "Silent Spring." "Silent Spring" in turn is represented as the largest unit of the chart, in my opinion the most important single study in ecosystematics that has been written. For other social units, the role of ecology gives many surprises. It is the dominant factor in the White House, but only the thinnest of lines in federal and state agriculture, and even in the pest-control-philosophy-dominated National Academy of Sciences. Natural resource and conservation groups have shown unusual success in receiving, and giving out, sound ecological information. Our campuses make a poor showing beside them, for although much ecologic knowledge goes in to academia, but little goes out (to wit, the "silent scientists"). The role of "ecology" as a social unit in America is unquestionably the most illuminating feature of the entire chart. Whereas this science should be playing a leading role on the social scene, it is represented as the smallest of social bodies which, altho sound knowledge goes in, so little comes out to influence the public that that little is indicated only by a dotted line. (Figures 1,2 and 3 were prepared by Harry E. Van Deusen, Research Associate, Aton Forest.)

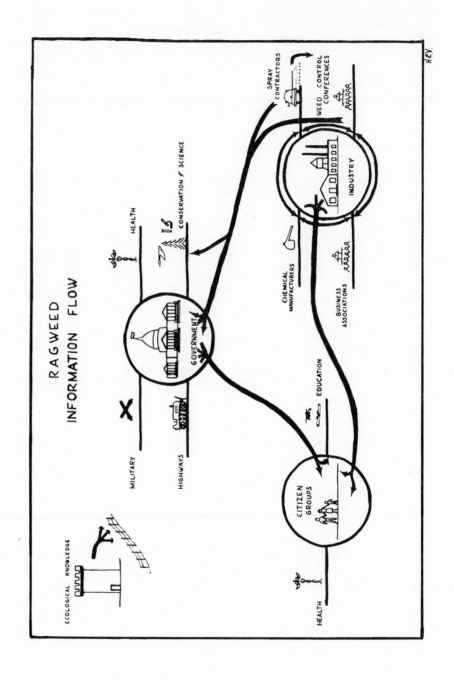

Fig. 2. *Flow of Roadside Ragweed Control Knowledge*. Chart based on a questionnaire survey made in 1958–1959, to over 200 social units involving two federal governments, seven states and provinces, five counties, 2 major cities, and miscellaneous individuals and organizations. Each social unit was asked from whom they did get or would get information, what literature they had published or released, and to whom it was distributed. The social units recognized were of three major categories: government (including military, highways, health, and conservation & science); citizen groups (including health and education); and industry (including chemical manufacturers, business associations, spray contractors, and weed control conferences). Two kinds of knowledge were segregated. The first was *ecological* (blue-line)* knowledge, which recognizes that ragweed (Ambrosia artemisiifolia) is a pioneer plant of sterile sites, the "symptom" of a "site disease." It will disappear if the site itself is altered; but will become more abundant if its competitors are destroyed by spraying. This knowledge is not communicated to government, industry, or citizen groups. The second type of knowledge is that associated with commercial herbicide blanket-spraying (red-line)* which does kill ragweed but also kills its competitors, so opening the site to more ragweed and to grassy weeds. Separate work sheets were prepared for the flow of information in each of the states, provinces, and federal governments. The present chart is a generalized diagram abstracted from these worksheets, and includes only the more important flow lines. It will be noticed that the basic ecologic knowledge (upper left of chart) is restricted to its source, and is not communicating itself to the rest of society. Commercial blanket-spray knowledge (rest of chart) appears to arise in industry, though even there the knowledge goes round and round and round and there is no definitive source. There is a heavy flow-line out of the spray contractors and the weed control conferences (and from the contractors *to* the weed control conferences) to government, especially to conservation and science. This same information in turn is communicated to the citizen groups, reinforced by heavy communication directly from industry.

* Blue-line indicated by white lines; red-line by black lines (See Fig. 3).

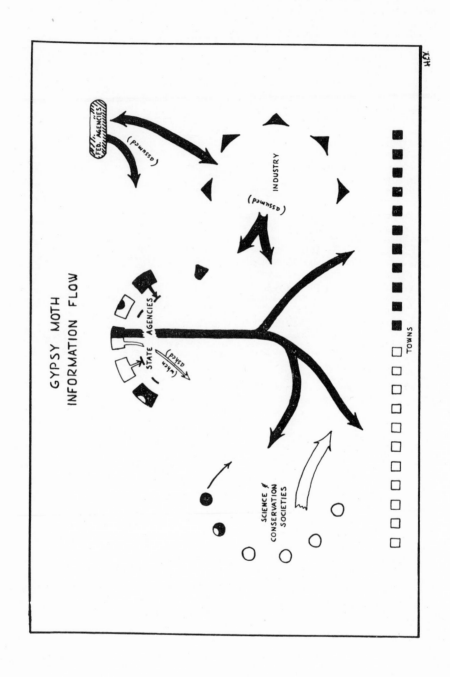

FIG. 3. *Flow of Gypsy Moth Control Knowledge.* Chart based on a questionnaire survey made in 1962, of approximately 100 social units in a single New England state. Each social unit was queried as to its publications if any, as to whether they recommended or wanted broadcast aerial spraying with DDT of large acreages, or wished alternative chemical or biological controls in local spots of high infestation or research directed to such alternatives. Other values of the human ecosystem were considered, involving timber, wildlife, water supplies, and human health. Each social unit was also queried as to from where, and to whom its information was communicated. The social units recognized were of five major categories: state government agencies, federal government agencies (when mentioned by others), science and conservation citizen societies, industry (when mentioned by others), and the various Towns themselves (which voted on aerial spraying). Two kinds of knowledge were segregated: ecological (blue-line) knowledge represented on the chart by white lines, opposed to indiscriminate broadcast aerial spraying with DDT; and red-line knowledge represented on the chart by black lines, favoring indiscriminate broadcast aerial spraying with DDT. The present chart is a generalized diagram abstracted from the questionnaire returns and interviews, and includes only the more important flow lines. All except two of the reporting science and conservation units (symbolically represented by six units on the chart) were blue-line in their knowledge, and the two red-line exceptions clearly showed the influence of one state agency on people simply unable to judge the ecological nature of the problem. The Towns are shown equally divided between blue- and red-orientation. (The exact number of each is not given, so as to preserve the anonymity of the state involved). Only one Town made an intelligent effort to obtain expert advice from national authorities; it voted against aerial spraying. Voting in other Towns was rarely significant, since the die was often cast by a handful of uninformed emotional or selfish citizens, influenced only by a pro-spraying representative. "Industry" and "Federal Agencies" are included in the diagram. Although they were not directly queried, they were frequently referred to by others. Furthermore, since there was no actual scientific source of information in the state, it must be assumed that pro-spraying "knowledge" arose either in industry, or federal agencies, or both. The situation with the State Agencies was clearly the most revealing and significant part of this sociological study! The agencies queried were those involved with agricultural research teaching and extension, with fisheries, game, forests, parks, and health. The "official line" from each agency, without exception, was that they make no recommendations whatever concerning aerial spraying. This was empirically true of four of the agencies. These four, despite the soundness of their ecologic knowledge, or the enthusiasm of their unecologic knowledge, did what they wanted; and kept mum. It was the fifth state agency however which eventually supplied the secret to the entire problem in the state. This tax-supported agency, and essentially only one department in it, and essentially only the head of that department, appeared to be the one and only source of such pro-aerial-spraying information as led others to report that "the state" recommended such spraying. On the other hand, when in the presence of or queried by a blue-line citizen, this agency gave the winning impression of being 100% blue-line itself. Although such an agency must be referred to as being two-phased, its effect on the social scene was however 100% red-line.

3.   Science and Technology

Man can hardly recognize the devils of his own creation.

—ALBERT SCHWEITZER (1875-1965)

The relationship between science and technology is much more than the relationship between man and maid, far more even than between mad man and made maid. I would suggest that Science be likened to Athena, the divinity of Greek genius in all its intellectual and artistic aspects, the virgin goddess sprung from the head of Zeus, heir to the nobility of the human race.

Technology presents quite another face of the human race. First the seducer of Science, then the keeper of a mistress; then a pimp; the recruiter, to enlarge the stable of fillies; then the master behind the madam; and finally the supreme ignominy, permissiveness as to births, with strangulation of male births and careful nurture raising and training of female children for future enlargement of the farm and milking barns. Never have so many been at the mercy of so few. The 20th century explosion of knowledge has been largely technologic, not scientific. This technology now calls itself Science!

"Technology . . . Never have so many been at the mercy of so few."

I am not opposed to technology. Every man has an urge to ride a filly, an urge which is frequently transverted satisfactorily to other objects and to the highest artistic ideals of the race. Technology, as the practical application of basic scientific knowledge, has given us better food, better shelter in terms of heat, clothing, housing, better health and longer life, the "best" life that man as a species has ever known. It has given us the telephone, radio, television, books, music records, highways, bridges, aviation, hot showers, and flush toilets. It has amplified the effectiveness of the output of human energy so that one gentle touch of a button can put into operation an automated factory producing a "needed" object—the perfect materialization of the Sorcerer's Apprentice. It would also explode the atomic bomb that could destroy a city. Thus, we see that an infinitesimally minute *in*put of human energy can result in an ever-increasing *out*put of total energy. This *Sorcerer's Apprentice Principle* involving the amplification of energy, is giving us ever-increasing leisure (or emptiness) in life; or it can destroy us.

We referred to the *Sorcerer's Apprentice Principle* as one involving the amplification of human-oriented e n e r g y. A minute input by man, as by the pressing of a button, effects an enormous output, as by the explosion of a hydrogen bomb. Let us now extend this Principle with the thought that it involves the amplification of human-oriented m a t t e r, or things best known as gadgets—the true symbol of the 20th century. Regardless of how we interpret our innate biologic compulsions, and how we evaluate their interractions with a social "system" (in the holistic sense) that thrives on ever-increasing economic production and consumption, the pragmatic effect is that mankind is engaged in a mass mania first of producing, then of buying manufactured articles we do not need, at prices we cannot pay, on terms we cannot meet, because of advertising we do not believe, to attain a status we do not respect, amongst people we do not like.

In the process, more and more of our daily existence is spent between unhappy time on a job which produces more gadgets, and unhappy time servicing an over-begadgeted home, divided by unhappy time commuting between these two monsters. Periodically there is unhappy time on the escapes that are called "vacation." Humorously, such a life is rationalized as one of "saving work" in order to have "free time" and "leisure." Gadgetry is godhood, and the services at her altars and the maintenance of her temples are demanding more of our sacrificed lives than any god in all of human history. This is Technology, once the servant of Science and useful for interludal pleasures, once mistress *to* the master, and now Machiavellian mistress *of* the master.

One example should suffice to indicate that as the second half of the 20th century progresses, Technology has become Science. The star in the crown of a most highly successful series of books on science for the layman, a series that is lavishly illustrated, heavily researched, with no expense withheld for corralling the most eminent authors and editors, is entitled THE SCIENTIST. Titles of the chapters set the stage: "Hero—and Human Being," "A Landscape of Poetic Vision," "A New Elite," "The Instruments of Conquest." "Expanding Realm," "Accolade for Greatness." Athena, you say?

The first four chapters, the first half of the book, are ostensibly on science and the scientist, altho it takes n o great perspicacity to see that the practical, the engineering, and the technologic are motivating forces in the discussions. Like a stage drama building up to a melodramatic climax, the sixth chapter is entitled, "A Booming Establishment." The example of the big boom is no canon of a cathedral, but the heavy concentration of university scientists in one part of the country, supported by encouraged by and presumably bedded by an equally big booming aggregation of surrounding technology-oriented industry. The seventh chapter is the climax of this eulogy of The Scientist—entitled "The Bounty of Technology." You can imagine how the bounty booms. A recessional to close this imposing liturgical mass is entitled "Accolade for Greatness." The accolade for greatness is, of course, the Nobel Prize, worthy of Athena herself if isolated from this present context.

We must add that the Nobel Prize is given for creativity in the fields of physics, chemistry, what we now call molecular biology, and medicine. The ecological fields that involve the concept of man-and-his-total-environment are notable by their absence! The world greatly needs a Nobel prize in ecology.

 "Technology—Herein lies the irrationality of our rationality."

I am all for Technology—if Athena remains mistress of the mind, and not of the mat. Yet should we not ask ourselves whether technology has gone far beyond its competence, whether technology has not become mankind's major religion, with its god demanding ever-increasing human sacrifices—highway fatalities, plane crashes, thalidomide babies, urban escapees (as moths swarm to flames) leading to drug escapees, not to mention the low-level chronic effects of an all-pervasive environment loaded with metallic poisons and organic biocides. Herein lies the irrationality of our rationality.

> *The rebels are right in being pessimistic . . . I do not think they are even pessimistic enough. To me it seems possible that the new amount of technological power let loose in an overcrowded world may overload any system we may devise for its control; the possibility of a complete and apocalyptic end of civilization cannot be dismissed as a morbid fantasy.*
>
> —Don K. Price
>
> *(in Amer. Assoc. for the Advancement of Science presidential address, Science 163(3862):25-31, Jan. 3, 1969)*

4.   The Value Judgment

Since scientists do not judge the laws of nature, a social scientist cannot judge the laws of his society. I once asked a great expert on the American system whether the decision of the United States Supreme Court desegregating the schools was good or bad. He replied, "As a social scientist I do not make value judgments." He indicated that he had personal, unscientific views about the question I had asked but he exhibited no confidence in them, appearing to think that they were the accidental, and indefensible, product of his early environment.

—ROBERT MAYNARD HUTCHINS
Doing What Comes Scientifically. Science began as a search for knowledge. It is becoming part of the search for power. (In The Center Magazine 2(1). January, 1969)

There is one strange and irrational emergence from the contemporary cult of the descriptive and objective, of the quantitative and empirical, of the worship of the technical and denial of the ethical. That emergence has created one of the most pathetically humorous and psychopathically ludicrous viewpoints in the history of science, as well as in the history of the human race.

I refer to the bigoted antipathy on the part of scientists (i.e., technicians) for what they contemptuously call "the value judgment." Not only does this cerebral disease infect most scientists. It has spread to other fields, to the humanities and the arts, to all facets of liberal education where the cult of the quantitative and the capacity of the computer permit mediocrity to establish a mass-oid methodology.

The term Value Judgment needs no explanation. It is a decision, an opinion, a judgment based on the human thought of an individual. It is a judgment outside the confines of a designed methodology for a specific little research project; it is a judgment as to the reality and good-ness of that little situation in terms of a larger and more comprehensive ecosystem. It was the judgment of Rachel Carson in 1962 when she stated that altho DDT killed, for that year, the malaria mosquito, the gypsy moth of forest trees, and the cotton boll weevil, it also killed the parasites and predators of those insects, and moved thru the water soil and air, dispersing itself in the entire earth ecosystem, to accumulate many years later in the food chains of organisms including man, to affect normal behavior and even to result in death.

Perhaps the most illuminating example of the cancerous hypertrophy of this intellectual disease revealed itself during a five-day conference in 1967, sponsored jointly by a union of ten leading liberal arts colleges, by an eminent national institution, by an industrial "educational foundation" and by a private local foundation active in conservation and natural history. It was attended by 45 supposedly handpicked administrators and professors from the liberal arts colleges, representing all the major areas of education in those institutions—sciences, arts and humanities. The purpose of the conference was to explore the possibilities of integrating the courses and departments of these colleges into an awareness that they were not isolated little cubby-hole specialties, but parts of a single humanitarian-oriented Total Ecology that comprised the unity of man-and-his-total environment. Of the local foundation—that had much to gain by this conference—its President went off on vacation; one member of the Board of Directors and one professional staff member attended the opening lecture. The opening lecture and field trip, by an ecologist of unusual parts, was intended to set the tone and tenor of the conference. It was met by derision and non-understanding on the part of the conferees, and altho that lecturer attended all succeeding meetings to be available if wanted (his own little experiment) he was all but totally ignored. The other scientist, of international eminence in his field, and one of the three major individuals of the Conference, sat essentially unutilized (part of his experiment). The various meetings were interesting. They reminded me of the beer-and-bull sessions of my sophomore days. Yet these meetings were dominated by aggressive activists reiterating time-worn platitudes, punctuated by contributions from a motley assortment of uneducated specialists with extreme prejudices and an absolute horror of a Value Judgment—the qualitative opinion of the other fellow, that is in opposition to your own quantitative truth.

The ironic climax to this five-day fiasco came during the very last meeting when the frayed tatters of the hopes of this conference were drawn together—not by one of the illiterate doctoral professors but by the only self-educated humanitarian in their midst, in such a way that one of their leaders was forced to admit that perhaps they had now reached the stage of thinking and understanding that they should have been at *before* they even came to the conference. (Subsequent developments however, to my knowledge, indicate no more than a rapid charge to an old storage battery: it froths at the moment, and is dead the next morning. The liberal arts colleges are still seeking "innovation" for the sake of innovation. The local foundation is circulating its activities in "education in natural science" by a leaflet with drawings that would aim to separate a fern from a mushroom, a songbird from a duck, and a butterfly from a bunny. Even Total Ecology at the eminent and sponsoring national institution seems in process of searching reevaluation.

The problem of the Value Judgment does not allow for solution in this age of 20th century science. It is the inevitable result of the intellectual mediocritization of the Scientific Establishment, with the dominance of technology and technicians, furthered by the spread of this worshipped philosophy not only to other fields of knowledge but to the general populace. It is part of the anti-intellectualism of the democratic age, a shameful abdication of human responsibility.

Bentley Glass

SCIENCE AND ETHICAL VALUES

Chapel Hill: University of North Carolina Press. 101 pp. 1965.

In the odd literature that clutters library shelves concerning the bonds between science and ethics, it is difficult to choose volumes of significance and lasting value. Most of these books are written by plodding scientists who also have a smattering of social consciousness. Frequently these men, with the ego of their declining years inflated by the praise indiscriminately lavished upon retirees whom society can at last put on the shelf, frequently produce grandiose philosophical balloons, that die with a wheeze before they do.

From this welter, it is a real pleasure to recommend this small book by Bentley Glass, whose scientific speciality is human genetics, and who has been active in a wide variety of administrative responsibilities, in research and in education at both high school and university levels.

The book is in the form of three lectures. The first chapter is entitled "The evolution of values." Within the framework of genetics, Dr. Glass traces the development of values, starting at the molecular (Third) Level of Integration, and progressing thru the Fourth, Fifth (the individual), Sixth (the population), Seventh (the community) and Eighth (the biome). The Ninth Level is not formally recognized, but the germ of the thought is inherent in all these pages. The conflicts between the morals at different levels, even the value of death itself, are rationally placed in the light of relative good and evil.

The second chapter "Human heredity and the ethics of tomorrow" is an exploration of our moral obligations—not for the immediacy of today—but for the well-being of humanity in the future. In terms of recessive, detrimental and lethal genes, genetic diversity, gene flow between groups, undesirable mutations from various sources, he asks whether we have the right to inflict genetic damage on our neighbors, and on future generations. All for a present short-term benefit.

The third chapter is a penetrating discussion of "The ethical basis of science." "As long as science is a *human* activity, carried on by individual men and groups of men, it must at bottom remain inescapably subjective." Accepting this inevitable subjectivity of science, he moves on to the two-world belief of most people: the positivist world of *is,* and the moralist of *ought.* He spotlights Bronowski's magnificent thought that scientists *"ought to act in such a way that what is* true can be verified to be so." "By examining critically the nature, origins, and methods of science we may logically arrive at a conclusion that science is ineluctably involved in questions of values, is inescapably committed to standards of right or wrong, and unavoidably moves in the large toward social aims." After further discussion of obligations, commandments and responsibilities, Bentley Glass closes with the words "Science is not only to know, it is to do, and in the doing it has found its soul."

The problem of science is not the vision and action of such as this man, but that there are so few like him in science.

5.   Science and Culture

By "culture" I am not thinking of the bacteriologist who cultures
his germs in test tubes and petrie dishes, or the farmer who cultures
the vine in the open fields, or the cartographer who symbolizes the
man-made cultural features on his maps, or the archeologist who describes the culture of a primitive tribe, or even the historian who admires the ancient culture of the Orient. I am certainly not referring
to the consciously cultured snobs who have acquired a veneer-eal patois
of the thinnest kind, and spray their misty droplets—as others would
when they cough—into every social gathering. The culture which I
have in mind is in many respects incapable of definition. It is a mysteri-
ous human creation, in many respects antithetic to the Technology
described above, and often profoundly menaced by it. Culture implies
what is desirable and ideal in the flowering of humanity, whether it
is for the individual, or the civilization of which he is a part. It implies
an enlightenment and refinement of taste, a development of potentialities by training in physical, intellectual, logical, esthetic and moral
spheres. It involves the humanities, the fine arts, a n d science. Too
often, culture in this sense is associated exclusively with music, literature, poetry and painting, with philosophy and religion. It need not
be so limited.

I do not, however, wish to imply that a Ph.D. in science will auto-
matically add to one's cultural attainment. Indeed, I would say that
there is a negative correlation between certain attributes of the Ph.D.
in science, and the cultured individual. Further, the doctored syco-
phants of science have become the worst enemies of the humanities.
They have spread within the traditional humanities a crippling inferiority complex that has led to a loss of confidence in dealing with
qualitative problems of value, taste, and belief. One wonders whether
the humanistic heartbeat has failed. This scientist of Academe is not
only trained as a narrow specialist, but the narrower he is, the better
scientist he is (as a publishing researcher). I am not, however, sure it
is the training which makes him that way: he enters these fields in the
first place because of aptitudes that favor isolation and narrowness; he
admires the "reality," the "truth" and the pseudo-certainty that are
the hallmarks of the technician and the technologist. For culture, I
would any day choose a self-educated peasant rather than a Ph.D.

It is my thesis at this point, nevertheless, to state emphatically that
science is not antithetic to culture, as many would maintain. Such
detractors of science see only the low-brow contributions of the technicians and the technologists. (Indeed, the technologists do unwittingly
add to culture, as I some day hope to document in full length.) It is
my thesis to state emphatically that science is an ingredient *of* culture,
if not the greatest ingredient. No mere assertions, no lengthy oratory,

no persuasive rationalizations will convert the anti-science doubters, of that I am well aware.—*I merely submit the entire knowledge of science itself,* in all its voluminous magnificence. Granted that much of it is in the language intelligible only to the high priests themselves, indulging in the pleasures of their own society, an increasing amount of it nevertheless is being translated into language for the layman.

In so evaluating the relationship of science to culture, we must not neglect the newer sciences of man himself. They are the last to be developed in the history of the human race. We have looked with relative objectivity upon the rest of Nature. It is most difficult, however, to be objective about ourselves, and to give up the idea that man is the pinnacle of pure intelligence, and that Nature was made to be exploited by man.

To know and to understand man and the world about him is to admire, to love, to forgive—and finally to live in order to improve the very values which are the essence of culture itself.

* * * * * *

I have sought to portray science not as it "ought" to be, not as it "should" be, not as it "might" be, but as it is. I have sought to portray the actuality of today, not the possibility of tomorrow. The science of the scientists is a closed and complete philosophy, such as has characterized every dogmatic theology thru written history. "This sort of completeness is like the conceptual framework of a prescientific culture, in which all the questions that may be asked may also be answered," as said by philosopher R.G.A. Dolby in *Science* 159 (3815), Feb. 9, 1968, "without any need to observe phenomena more carefully under a different context. Indeed, *only* those questions are asked which *can* be answered." In this approach, the whole system can never be found to be wrong. It itself is a growing stable self-contained homeostatic "system."

One should have a sniveling suspicion that any science which *is* considered so complete in its philosophy must have something wrong with it? Ah, no! No dogmatic theology admits to incompleteness. No such admission is part of the rules of the game.

* * * * * *

I disagree with this view of Science as necessarily a closed and complete theology, closed and complete only because of the intellectual limitations of its high priests and the slavish acquiescence of the congregations. To the contrary, it is the openness of Science—and Science alone—which gives it its special and valuable characteristics. It is the openness of Science which will allow for a growth, an unfolding, a flowering, a building upon the best of the past, to attain to new dimensions of grandeur for the human race and for the space ship on which it is traveling. The stage of our discussion is now reset. It is no longer the actuality of where we are, but the possibility of to where we may go, that now will claim our attention.

And he said unto them:

BE FRUITFUL, AND MULTIPLY, AND REPLENISH THE EARTH AND SUBDUE IT; AND HAVE DOMINION OVER THE FISH OF THE SEA; AND OVER THE FOWL OF THE AIR, AND OVER EVERY LIVING THING THAT MOVETH UPON THE EARTH. AND GOD SAID, BEHOLD, I HAVE GIVEN YOU EVERY HERB BEARING SEED, WHICH IS UPON THE FACE OF ALL THE EARTH, AND EVERY TREE, IN THE WHICH IS THE FRUIT OF A TREE YIELDING SEED; TO YOU IT SHALL BE FOR MEAT. AND TO EVERY BEAST OF THE EARTH AND TO EVERY FOWL OF THE AIR, AND TO EVERY-THING THAT CREEPETH UPON THE EARTH, WHEREIN THERE IS LIFE, I HAVE GIVEN EVERY GREEN HERB FOR MEAT: AND IT WAS SO.

Where TO?

THE POSSIBILITIES OF TOMORROW*

*It is planned to elaborate the material of this section (pp. 133-end) in another volume, the second of a trilogy, of which the present volume is the first.

THE NEED: A NEW SCIENCE

How does the human race propose to survive and, if possible, improve the lot and the intrinsic quality of its individual members? Do we propose to live on this planet in symbiotic harmony with our environment? Or, preferring to be wantonly stupid, shall we choose to live like murderous and suicidal parasites that kill their host and so destroy themselves?
—ALDOUS HUXLEY (1894-1963)
The Politics of Population

Our survey of Science as to the actualities of today, as to where we a r e, forces us to take certain stands that are at variance with those of most, but not all, scientists. We find that science is not the Open Sesame to "truth" and "reality" that many would like it to be. It has all the characteristics of a closed fundamentalist philosophy or theology, with its dogma, rituals, prophets, disciples and followers. As such, it is the most recent in a long series of scholastic beliefs and traditions that have dominated the thinking and behavior of man through history. *Scientism* is the better word for this ritualized body of thought.

Scientism is failing mankind. The 1960's and 1970's are going down in history as the time when certain lone voices of previous decades finally began to make themselves heard. Large segments of humanity are realizing that the future of the Earth is jeopardized. This future involves more than man and the factors of his environment. It involves the integrated single web of man-and-his-total-environment.

We look about us. Unless we are blind with prejudice, or simply blind, we view a situation that is totally inconsistent with any pious pronouncement about the "intelligence" of Homo sapiens. We find an exploding human population that will soon reach a simple volumetric impossibility. A technology is linked with a belief in an ever-expanding economy—an "affluence" for everyone—that in turn drains our natural resources, pollutes our environment, and cannot even be logistically extended to all the populations now living.

While we are myopically concerned with the short-term problems of food and health, the basic nature and needs of our biologic heritage are ignored or frustrated. While we concentrate ourselves into urban ghettoes under conditions totally alien to our ancestry, the noise of that environment and the disregard of our territorial imperatives and of our need for space, lead to escape by drugs on the one hand and by mob violence on the other hand. At the national levels, our most all-consuming activity in terms of money and energy is military (man against man) activity, and military preparedness of such awesomeness that one button accidentally pressed could result in the destruction of life on Earth to the point where man as a species could not survive.

Man only lives in relation to his Total Environment. And as we look at the Total Environment we are so shocked that we talk not only of "the pollution of the environment," not only of "the quality of the environment," but of "r e s t o r i n g the quality of the environment." There is increasing waste per person tossed into the environment (waste is good for a growing economy), increasing pollution of water air and land (we foul our nest), increasing uglification (from the sterility of the contemporary arts to the identification of "art" with that ugliness, as when an eminent art director collects fenders of wrecked cars, hangs them in a heap on his living room wall, and calls it "sculpture"). It is not necessary here to document the pollution decay and destruction of the Total Environment, and its impoverishment as an environment for man.

Whatever our solutions for specific problems, in the last analysis it is the *number* of men which will determine the *survival* of man. If you have more than two kids, you are part of the problem. If you have two or less, you are part of the solution. If the compulsion to breed is stronger than the compulsion to live, we are headed for extinction. The subject is fully discussed in a long series of excellent books.

Scientism has been weighed in the balances, and found wanting. The evidence in ourselves and around us is clear and indisputable. We need a new philosophy, a new philosophy of science, a new science. Before however that philosophy can be intelligently discussed, we have to gain a clearer idea of the way scientists have already divided up their "domains," of the way their Ivory Tower has been cut up into apartments and rooms and cubicles, each with its shelves and cabinets, their drawers and trays. In short, we need an understanding of what might be called a "classification of the sciences." There are many classifications of the sciences; most of them can be likened to cutting a pie into pieces, a simple arithmetic separation into parts. No such simple classification is suited to our needs. The complex of relationships between the existing sciences needs a model of higher intellectual caliber. The next three chapters (pp. 119-137) will present an interpretation of the situation amongst the pertinent biologic sciences as it empirically d o e s exist—not as it should exist according to some apriori theory. It will be upon this frame or model that the final proposition of this book will be constructed.

A. DEVELOPMENTS FROM CLASSICAL BIOLOGY

I choose to start my discussion with the specialized field of biology, the study of organisms. In history, this is where much of science did start. After all, man himself is a biologic blob, albeit heavy-headed and thin-skinned and far inferior to other organisms in many of his direct perceptions of his environment. The biology of the first half of the 20th century was largely "organism" oriented. The student studied the earthworm, the crayfish, the frog, the shark, the rat—and man, as different kinds of organisms. Other non-organism aspects of the world about us—morphologic, physiologic, historical, geographical—were not ignored, but they proved conceptual efforts as we tried to overcome our organism-interpretation of the world about us. Such inhuman phenomena as sponges, bee colonies, and plant-communities gave us no end of trouble as we likened them, and identified them (and still do) to individual organisms such as ourselves.

At the turn of the century, two developments from Classical Biology began to take form. One "leftwards," engaged in smaller and smaller components of organisms, became known as Molecular Biology. Utilizing the methodologies of chemistry and physics, (the vaunted "exact sciences") it is associated with some remarkable breakthroughs in our knowledge and comprehension of its multiplex minutiae. The other, "rightwards," dealt with larger and larger "wholes" of ever-increasing complexity, and less and less amenable to the precise methodologies of the crowd-phenomena of minutiae. Here are the Environmental Biologies and the Environmental Sciences that are struggling for recognition as the 20th century wanes.

		(Ecology)
MOLECULAR	**CLASSICAL**	**ENVIRONMENTAL**
Biology	**Biology**	**Biology**
(BSCS Blue)	(BSCS Yellow)	(BSCS Green)

In high-school teaching, this three-fold segregation received a definitive birth in 1963 with the publication of the three Biological Sciences Curriculum Study textbooks. The BSCS program was born from the combined efforts of professional biologists, operating at the start thru the American Institute of Biological Sciences, and high school teachers. The preparation leading to these books went thru several years and four million dollars of summer workshops, conferences, and trial teaching. It is upon the ecology-oriented Green Version that I would like to place a spotlight. For one thing, it is largely the writing of one man, Marston Bates, and it bears the imprint of his personality and his style, which the numerous editors and committees had the good sense not to destroy. Despite certain defects (not of this man's authorship), it is the finest ecological work of its kind, which I would unhesitatingly use at this time, at the undergraduate college level. Eventually, as with all science books, it will be superseded. When so, I trust its successors will retain its values. This book will be extremely useful in the hands of high school teachers who by their lack of specialization, by their close contact with students, parents, and the general citizens, by the very nature of their personalities, are the most effective communicators of ecological thinking.

In colleges and universities we found—understandably—a very different situation in the later 1960's. In universities, the emphasis is on research, not on teaching, so much so that many universities are in fact if not in the catalogs, research factories. Molecular Biology is admirably adapted to this explosion, for it can operate on short-term projects, and turns out papers that impress the chemists and physicists, leading in turn to ever-more-effective grantsmanship. I do not say this disparagingly, but as a sociologic observation. The result of this development is that deans and administrators have gone on an orgy of forced integration of botany and zoology departments. More often than not the classical botanists and zoologists (each with their own and markedly distinctive personality types, and still having logical and rightful places in research and education) have been left like flotsam and jetsam in the older buildings, while the Molecular Biologists rise ever higher in tall expensive buildings, often shared with such "exact" sciences as chemistry and physics. In the meantime, the poor "ecologists" (those who had taught but one now-woefully-outmoded course in either a botany or a zoology department), if they are alive at all, had to choose between their completely obsolescent "branch" of either botany or zoology, or grow into a newer (Green Version) ecosystem-oriented ecology. This newer ecology is not only at war with the in-power self-adulating Molecular Biologists of their departments, but is pathetically out of place in the rigid loyalty-dominated department structure of our universities, and even with the administrative conventions that cleave a campus into "colleges," colleges of agriculture, forestry, arts and sciences, engineering, and the professions. I see no solution to this very real problem of the university, even if more deans become ecology-oriented (as I know deans, this would be an unanticipated accident) , or until Presidents are ecologists or ecologists are Presidents. Let the ecologic revolution come.

In this impasse of the university, I look to the small liberal arts college as the more promising. The liberal arts college has its flexibilities, its orientation to liberal (if no longer terminal) education, its relative openness to innovation, to point the way that some day the large universities may follow.

> *". . . no solution . . . until Presidents are ecologists or ecologists are Presidents."*

It is realistic to remark that each of the three major fields here discussed—Molecular Biology, Classical Biology, and Environmental Biology—attract unto themselves specific and distinctive personality-types. It is not their respective and different educations which have made them different from each other. I would say they showed different aptitudes from the start, both by inheritance and by childhood environment. Get three such people together some day, and listen to them discuss the eternal verities, and watch them—the way they think, their interests, mannerisms, appearances. It is very much like bringing together three such mammals as a seal, an ape and a bat, for a conference on mammalogy. Could it be done, I would predict they would not discuss their common mammalian traits. None would ape another, and each would seal his arguments with batty claims.

B.   THE NINE LEVELS OF INTEGRATION

During the latter half of the 19th century, the dominant unifying concept in all of biology was that of *e v o l u t i o n* . Thru the system of binomial nomenclature, as initiated even earlier by Linnaeus, a *species* became a recognizable unit in nature, a relatively homogeneous pause in the long-term change of biologic material. These concepts were admirably suited· for the development of traditional now-classical Yellow Version biology. They will always be suited, if such is our only goal. On the contrary, when biologic interests began to extend, as to "plant ecology" (better called plant sociology) , the deliberate carry-over of these classical ideas created some very foolish scientific methodologies, the impracticalities of which were seen by some scientists from their very inception.

There is nothing new about the concept of holistic *levels of integration*. It has been hanging around for a long time, particularly amongst the philosophers. It says no more than that the whole is greater than the sum of the parts, that H_2O is different from and cannot be predicated from the nature and characteristics of its parts, hydrogen and oxygen. It is sister to the idea of *emergent evolution,* the concept that looks for the evolution of new emergent qualities when parts combine in nature to form wholes. Offbeat and beatnik scientists of past years repeatedly tried to bring it to the attention of the in-power scientific world. As long as scientists were chiefly involved however with species and their evolutionary status, the idea was not only uninteresting, it was heretical and even mystical, with its implications of new emerging phenomena that could not be made part of precise equations. With the ecological departures from classical biology, however, we are beginning to look upon the world as wholes within wholes within wholes, without the sharp boundaries that separate one man from another. At least, in the mid-1960's, those scientists who have agreed to call each other eminent are beginning to use the term Levels of Integration.

For our purposes, we will begin with the "smallest" wholes in nature, and work "up" to the largest. We could equally well begin in the familiar "middle" (classical biology) and work to both directions, or with the largest (large ecosystems like lakes, continents) and work "down" to the smallest. I prefer the last as the most logical, but I use this approach only for children, since they are not already indoctrinated, by the misleading emphases of our educational system. There will be 9 Levels in this discussion, altho the story could be told equally well with more or less. Nature is not concerned with an exact number of Levels—only man is.

The First Level to be mentioned is that of the *subatomic unit,* the particles of the once-indivisible atom. These are the protons, electrons neutrons, mesons, and a swarm of other elusive little sprites that are tumbling out of some fantastically thrilling 20th century research, and revealing wholly new symmetries and architectures in the world about us. I once made the mistake of asking a physicist whether he thought he and his colleagues would soon reach the end of these discoveries. "Oh no," was his cheery reply, "if we cannot find others, we will start dividing the ones we have." Sure enough, two physicists dreamed up a particle of a particle, whimsically called it a quark. And soon the great quark hunt was launched.

The Second Level involves those wholes that are composed of various combinations of the subatomic particles. These are the now-familiar *atoms,* which are classified into a 100 and more *elements,* such as carbon, hydrogen (the simplest and lightest), oxygen, nitrogen, sulfur, such metals as copper and zinc, up to such complex and heavy atoms as lead and uranium. Here is the realm of the physicist and chemist, while changes and metamorphoses between and amongst the atomic particles moved from science to technology on that awesome day over Nagasaki in 1945. We now have enough potential airborne overkill in the atmosphere as would—if inadvertently triggered—bring to an end not only man as a species, but possibly this geologic age. If this is a mark of man's "sapience," my understanding of that word is erroneous. If so, I soberly suggest that next time animals do without it—sapience.

The Third Level comes into focus when we consider the combinations of atoms into *molecules,* of hydrogen and oxygen into water, of salt, sugar, chlorophyll, DDT, and DNA. The chemists, the biochemists, and the molecular biologists play in these fields, building up, breaking down, moving energy about, and in the process altering, improving or polluting not only our total environment but—sooner or later—man himself.

The Fourth Level—in the biology sequence—revolves around the idea of *cells,* highly complex aggregations of many different kinds of organic and inorganic molecules, organized, alive, acting either as independent organisms like an amoeba, or with "lives" of their own within the body of a complex multicellular organism. *Cytology* is a science with both a past and a future.

The Fifth Level, that of *multicellular organisms,* is the comfortable and familiar ground of traditional *biology,* involving the well-known plants and animals, the traditional courses of biology departments in morphology, physiology and taxonomy, and to a lesser extent, the geography and history of plants and animals (altho such subjects will be often assigned to geography and history departments, where they receive some queerly unique emphases). Even in this traditional field, there have recently been amazing new breakthroughs in the phenomena of animal behavior, otherwise known as ethology. No longer are we dominated by a rigid dualism, relegating behavior to the black or white philosophy of "Is it due to inheritance? Or to environment?", and fearful of the mortal sin of anthropomorphism which is the so-called error of interpreting animal behavior in terms of human "reasoning" intelligence and emotions. To the contrary, some of us are seeing a saner, if not so arrogant, zoomorphism, whereby our own irrational behavior is becoming more understandable—and therefore perhaps controllable—in the light of animal intelligence and emotions.

The Sixth Level emerges as a science when we realize that what also operates in nature is not just an individual bee or fish or goose or cow or man, but a swarm of bees in a hive, a school of fish, a flock of geese in V-shaped formation against the sky, a herd of cows, or a city of men. Be it the spirit of the hive, of the organized population behavior of mating male kobs in their East African putting greens, or the mob psychology of peasants in revolution against a governing aristocracy, the behavior of a *population* has become a field of study that is asking new questions, and opening new insights into our awareness of the world about us.

The sixth level is the first of those that bear the designation of "ecology." Levels six thru nine are collectively called "ecology"—

> *"Ecology . . . the only science
> for correctly assessing
> the negative aspects of technology."*

by some of those who think they have the correct usage of the word. Ecology, by any definition, is the only science for correctly assessing the negative aspects of technology.

At the Seventh Level different kinds of plants, different kinds of animals, or holistic combinations of plants and animals acting as wholes in nature, have been studied by a variety of scientists calling themselves *community ecologists*. Here belong the vegetation scientist, the phytosociologist, the forester, the range manager, and animal ecologists of various shades and hues. Nor are farms and cities of men exempt: The study of man-and-his-cultivated-plants (simplified and therefore unstable agricultural and horticultural communities) are to be considered at this level. Even urban man, with his cats, dogs, rats, mice, pigeons, canaries, goldfish, cockroaches, bedbugs, lice and disease organisms such as malaria and plague and their carriers, form highly interesting animal communities.

With the Eighth Level, we move into relatively unfamiliar ground, that of the *ecosystem,* and of *ecosystem science.* The word itself was coined only in 1935 by the British plant ecologist Tansley. He realized that under many circumstances it was not the plant community or the plant-animal community which was the studiable whole in nature, but the plant-animal-climate-soil-plus "whole" which formed an operable unit in the world about us, like a lake, or a coral atoll. Such a strange idea just did not fit into the Departmental organization of our universities, much less its applied corollary "conservation" (of natural resources) . Not only did it not fit, any attempt to establish it at once tended to jump, if not to undermine, the Maginot Line of offensive and defensive barriers between university Departments, thus bringing into action jealousies if not open warfare amongst these bastions of Academe. Before an ecosystem science—call it the new ecology, or environmental science if you wish—can become established in our universities, it will require not only scientific diplomats arising from within a department, but active and forceful administration not only from deans, but direct from the presidential office. Before Conservation can become established in our society, it will require more men of the caliber of Roland C. Clement.

It is well to remember that even tho the Ecosystem is the "largest" of the Levels of Integration in the sense of the variety of its components, it does not follow that the amount of space it occupies is necessarily large. The word ecosystem is a generic term. Specific ecosystems may vary in size from a moss-covered boulder in the forest, to a one-acre island, a huge inland area, an entire continent, or our planet itself.

I brought you into a plentiful country, to eat the fruit thereof and the goodness thereof; but when ye entered, ye defiled my land, and made mine heritage an abomination.

—JEREMIAH
Chapter 2 Verse 7

The Ninth and last Level is that of the Human Ecosystem. By human ecosystem we certainly do not mean a virgin, climax, primeval wilderness, which man has utilized, exploited, raped, or ruined, and which would return to its "balance of nature" if man would only "preserve" it. This is the archaic view of those scientists who are pegged at the Eighth Level, and have hit their intellectual ceiling. To the contrary, the idea of the Human Ecosystem is that *man-and-his-total-environment* form one single whole in nature that can be, should be, and will be studied in its totality. The implication is also inherent in the idea that one part cannot alter, move or change without all other parts readjusting. For this reason, the spiderweb is often used as an analogue of the ecosystem. All strands are interconnected in fantastic complexity. One cannot break, remove, add, or put a stress upon any strand but what there is an adjustment throughout the entire web.

"There will be a roll-back in human population."

A Human Ecosystem Science can hardly be said to exist. We sense the need for it, since our smart technologies are riding so far ahead of our basic understanding that we "do" first, and then mildly but with increasing worry watch the grim results as they unfold in 10, 20, or more years. Since we have but one world to play with, this mode of behavior is hardly an expression of intelligence. Because of the unplanned bastard-born technologic events since 1940, even our largest and most tradition-bound universities may be slowly setting up the administrative apparatus to handle both research and education in this subject. I refer particularly to the problems of those bombs called "dirty" (as tho none other are), which spread undesirable radiation throughout the one atmosphere of the one Earth. I refer also to persistent insecticides behaloed by short-term non-ecosystem-oriented agriculturists, as spurred by the chemical industry, who have distributed these biocides to the farthest ends of the oceans. Rachel Carson's *Silent Spring*—the first such ecosystem book in human history, and the work of a single woman undaunted by the human might arraigned against her—first alerted humanity to this danger. In addition and relatively suddenly we become aware we have dirtied our water, dirtied our soil, and dirtied our air—of which there is only a finite quantity on the earth—beyond a condition fit as a human environment. Our rocks and rills were once the hills from which rose our rivers teaming with fish and game. Now more likely than not, our rivers can be said to arise in the flush toilets of the exurbanites, and pass thru several human bodies before they are finally dumped, dirty, into the ocean.

The chief goal of a Human Ecosystem Science is a knowledge of, and a humanitarian-oriented technology towards, a permanent balance between man and his total environment—both operating as part of a single whole—that will afford a life of the highest quality. The quality of human population is at present out of adjustment with its environment. There will be a roll-back in human population. Either man or the rest of nature will take the initiative. If we have intelligence, now is the time to show it.

MOLECULAR Biology	←	**CLASSICAL** Biology	→	**ENVIRONMENTAL** Biology

Levels of Integration (Subject Matters)

← **Ecologies** →

1	2	3	4	5	6	7	8	9
Sub-atomic Particles	Elements	Molecules	Cells	Organisms	Populations	Bio-communities	Ecosystems	Human Ecosystems

In concluding this section, let us emphasize that the nine Levels of Integration are not water-tight compartments, either in nature, or should they be in our minds. The idea is simply a "model" that helps us to understand nature. Any one of the 9 wholes is but a part of the next larger whole, and is itself composed of parts which themselves are smaller wholes. Thus we have a boxes-within-boxes philosophy that gives us a totally different understanding than the divided apple pie which is exemplified by most university administrations. Such apple pies, or rather whole shelves of apple pies each to be cut in pieces, are the obsolete "models" that our present understanding of science is based upon.

I wish to add one further thought to this discussion of the ninth level of Human Ecosystem Science. In adding Man as a component of a larger whole, we bring into our orbit, as part of one and the same "system," all the man-centered fields of knowledge, not only medicine and law, political science and economics, psychology and sociology, but even the arts and humanities. Such a conceptual recognition by no means belittles those fields. To the contrary, Human Ecosystem Science aims to bring into relationship, as parts of one whole, not only the natural sciences but also knowledge of man himself. It is at this point that quality, judgment, value, art, and humanitarianism again enter into this one unity, as real parameters which—if we cannot intelligently manage—we can at least describe.

It is realistic to remark that each of the nine major Levels of Integration here discussed—sub-atomic particles, elements, molecules, cells, organisms, populations, bio-communities, ecosystems, and human ecosystems—attract unto themselves specific and distinctive personality-types. It is not their respective and different educations which have made these people different from each other. I would say they showed different aptitudes from the start, both by inheritance and by childhood environment. Get nine such people together some day, and listen to them discuss the eternal verities, and watch them—the way they think, their interests, mannerisms, appearances. It would be a study of prime importance for the student of Human Behavior. I would say that each has his own "intellectual territory," and he is neither interested in the territories of his colleagues, nor can he be convinced that they actually exist. Generally, one may be inclined grudgingly to admit the existence of a Level close to his own. Why not? It is either a part of his whole, or it is a whole of which his is a part. Yet of all the barriers between Levels, I would say that the greatest is between Number 8 and Number 9. The Number Eighter likes nature, not people, and is prone to study wilderness areas, virgin tracts, climax stands of forests, and recognizes man only as a polluter and a manager (who should not pollute and should not manage). The Number Niner likes people, and is generally a humanist. He is never more stupid than when he says, "I could not care *less* for open-space and the outdoors." And this kind of Number Niner is its greatest danger. He wants more food for more people with better health, but he cannot realize that these values are dependent upon clean air. clean water, clean soil and adequate space for each individual. Never try to make a Number Niner out of an Eighter, or an Eighter out of a Niner. They are basically different kinds of people, and each must be recruited from the common pools of the younger generations. The most that we can expect—and this is a function of general education —is that each accepts the existence of the other, and recognizes his importance in the unity of man-and-his-total-environment.

Marston Bates and Donald Abbott

IFALUK. PORTRAIT OF A CORAL ISLAND

London: Museum Press, 1959. 287 pp. 1958

Eighth Level science—the science of non-human natural eco-systems—is not enough. Although the elegance of its present systems analysis has puffed the pride of those quantitophilous scientists who can now hyper-quantify ten parameters (instead of the two or three of yesteryear), future historians may see the elegance closer to the inadequate simplicity of Moses' Ten Commandments than to the integrated knowledge of the future.

Development of a Ninth Level science will deal *not* with the "interrelationships" between man on the one end of a see-saw, and the Total Environment on the other end; it will deal with the *unity* that is man-and-environment. This "subject matter" is so infinitely more complicated than that of the Eighth Level that I have the gravest doubts but that the mind of man can only muddle along for many a millenium. Years ago, when I first began to play with the idea of a man-environment ecosystem, I was enthused with the thought that such an ecosystem science might be born and suckled on a coral reef. Here we have a man-environment system of the simplest possible kind—by comparison with those of continental areas. A reef is a biologico-geologic tropical formation, with its surface at or near sea-level. It is the boneyard of the organisms that created it, not only the coral animal, but countless other kinds of plants and animals. As the sea-bottom would sink thru geologic periods (over 4000 feet in one known instance), its surface would grow up to maintain its sea-level status. As the sea-level would fluctuate in recent Pleistocene times, waves would plane it down, or organisms would build it up.

A minor incident in the life of a reef are the piles of sand and rubble, swept in from the reef-edge by wind and wave, then dumped when that force was spent, to produce a coral islet. A ring of such islets, ephemeral events in time and space, create the coral atoll of fact and fiction. A speck of land bearing plants, animals and men, isolated in its own environment of air and sea, a knowledge of which would be ideal for the start of a man-environment ecosystem science.

Altho the roots of our coral knowledge go back to Darwin, and to many subsequent scientists, we are indebted to the Coral Atoll Program of the Pacific Science Board of the National Research Council. Especially are we indebted to its Atoll Research Bulletin (Number 1, September 1951; Number 118 and thereafter, published by the Tropical Biology Program of the Smithsonian Institution; Number 135, Aug. 1969) and to its indefatigable founder and editor F. Raymond Fosberg, for encouraging and organizing the wealth of knowledge that has come into existence, a wealth of specialized data, the stones and mortar of a new scientific edifice.

I wondered what scientist would find what atoll, from the meeting of which would emerge a neo-classic volume of the Ninth Level literature-to-be. Ifaluk found its Bates, as Galapagos found its Darwin.

Marston Bates is a zoologist of flawless reputation, bitten by mosquitoes before he was bitten by man. His style of writing, in books for the layman, has established him as a humanitarian, idealist, humorist, and one with infinite hope for the human race. In this particular book he has been ably assisted by zoologist Donald Abbott, who wrote the last nine of the 18 chapters.

Ifaluk: 260 people on a land surface of half a square mile. This speck of terrain in the Caroline Islands of Micronesia has been so isolated and unimportant that it has gone thru the days of domination by Spain, Germany, Japan, and now America, relatively unscathed. These people are "naked savages," living a communal existence, relatively untouched by the recent Industrial and the present Technologic Revolutions. There is no affluence or effluence; no poverty, violence, racism, educational problems, religious persecution, or discrimination about the blueness or whiteness of one's collar. Production and consumption do not figure. (Trade boats call at long intervals, to pick up copra, and to leave such items as iron adzes.) The insanity of urbanity is unknown, along with such highways and byways as those of Main, Wall, Park, Madison and a Gold Coast. Under a communal society, personal profit and all its attendant means and goals are simply non-existent. Strangest of all is the fact that the population is essentially stable. Illegitimate births are unknown, and the birth rate per couple is very low. (Just why is a tantalizing question that the authors ask, but do not answer.) This then is a society of naked savages, that we hope to civilize. They are described as kind, compassionate, intelligent, clean, healthy, happy, living in harmony with their environment and with themselves, guided by wise and intelligent chieftains. Work and play are so inextricably bound in their daily lives, as well as ample leisure, that our own compartmentalized time-clocked assembly lines, alternating with organized recreation, would seem a perversion. Here is a human society that restores one's faith in a potential goodness and worth of Homo sapiens.

That was in 1953. The authors are no starry-eyed romantic dreamers. They know that elsewhere in the Pacific islands were wars of territorial expansion, exploitation of the natives by their chiefs for the sandalwood trade, overthrow and collapse of a civilization as on Easter Island. They know that the health of the people of Ifaluk is related to the control of yaws by the American Navy. They know that man on an island is not man on a continent. Soon, if not already, an orbiting satellite will beam down upon this island the television programs of so-called civilization, inoculating the naked savages with wants and desires previously beyond their ken. Enter an entrepreneur to buy the island, for a gambling casino to serve the continental jet-set, promising employment for all the islanders to play-act their former lives for the bemused entertainment of the guests.

The Ifaluk of Bates, a relatively simple one-neighborhood man-environment system, is not only the opening call to a Ninth Level scientific literature, but clearly holds the secrets of some of the long-recognized ills of human society. Of the Ifaluk of the 1970's, I prefer not even to ask.

Gairdner B. Moment

BIOLOGICAL SCIENCE TODAY

Pp. 1-8, in Frontiers of Modern Biology,
Gairdner B. Moment, Coordinator.
Boston: Houghton Mifflin; 192 pp.; paperback. 1962

This book is subtitled "Twenty eminent biologists survey recent developments in talks first broadcast by the Voice of America." In 1961 the Voice of America cooperated with the American Institute of Biological Sciences in broadcasting this series of 20 lectures that was clearly aimed to reveal to the world the pinnacle of American biology. Many nations rebroadcast the lectures; other nations broadcast them in translation.

I most enthusiastically endorse this method of communicating to the general world public the knowledge of contemporary academic biology, American or otherwise. The success of this series only spotlights the more its deficiencies and failures. All the lectures are worthy of reading, and I recommend them, particularly that of Edward S. Deevey on "animal populations." Dr. Deevey was the only general ecologist in the lot, even tho he was assigned a specialist and restricted topic.

The volume is instructively fascinating in its limitations and restrictions. I assume the A.I.B.S. charged Dr. Moment with organizing the series, and had complete faith in his decisions of scope and content. Thus, my comments that follow reflect on the A.I.B.S. Dr. Moment is a zoologist, and of the 19 other "eminent biologists" he chose, only two are botanists. The others are in zoology or medicine, or in cytology biochemistry or biophysics. These latter are fields where one could not care less whether the whole organism is a plant or an animal. This is the way a zoologist typically views the world. Of the two botanists he did choose, both are specialists at the lower levels of integration. I just think it would be more honest to call the whole of the Moment "zoology" and not "biology."

Dr. Moment's own paper opens the volume and sets the stage. It is concerned entirely with Levels of Integration, the finest, most succinct statement of its kind I know in print. This is the first time to my knowledge that the concept of holism and of emergent evolution (altho he does not use those terms) is used as an integrating philosophy in a classification of the (biological) sciences. With precision and perception, he discusses the levels of: sub-atomic units, atoms, molecules, multicellular organisms, and populations of one species of organism (the Sixth) Level). *And then he stops!* Believe it or not, there is no mention of or comprehension of a level of plant-and-animal communities (which science has been around since 1900), or of the organism-climate-soil "ecosystem" (introduced in 1934 and now thriving), or of the man-plus-environment ecosystem (wherein lie all the problems of environmental deterioration and of overpopulation of human beings). This expression of American biology, displayed to all the world by the Voice of America, is a diamond-brilliant example of the way scientists of mid-century America have not risen to the academic and practical challenges thrust upon them by the cancerous growth of a Technology that threatens our very survival.

C.   THE 7 POINTS OF VIEW

There is nothing new about the idea of Points of View. It was presented in essentially the same form as given here by Tschulok, in 1910, and I am reasonably sure the model was then ancient, for it expresses —not necessarily nature—but the way the human mind perceives nature. Even tho we all do now accept the inter-convertibility of mass and energy, and the reciprocity of time and space, I am quite sure that none of you are likely to confuse the energy with the substance of your bread and wine, nor the time with the place of a rendezvous.

Let us assume that a specific "subject matter" at any Level of Integration be represented by a hexagonal pyramid. This subject matter may be an atomic particle, an atom, a molecule, a cell, a multicellular organism, a population of organisms, a bio-community, an ecosystem, or a human ecosystem. Whatever subject matter it is, either a concrete example of one, or a generalized abstract of many examples, we find invariably that our knowledge, our understanding, our management engineering or technology, tends to segregate itself into six Points of View. If we use the traditional and familiar biology as a case history, the nature of this concept should be clear.

The *first* question a child is likely to ask, and a scientist (for unless a scientist retains a child's curiosity, he is no good as a scientist) is "Of what is it composed?" "What is *inside* it?" This idea of *composition* is fundamental to the study of any subject matter be it mineral or a mountain, blood or bone. Here is the compulsion to "take apart" a watch to find what makes it tick, even if we cannot put it back together again. To understand a city however, or a nation, or the unity of our Earth, or even what makes one man tick, these things science does not yet comprehend.

A *second* question that is frequently asked is "What is its size and shape?" Here belong all the traditional *morphologies* (*morpho,* form or structure), involving a static descriptive approach to all subject matters. Size and shape has always interested man. I need only mention such a term as "36-24-36" and immediately a very definite size and shape comes to mind.

A *third* question, and bearing a pair-relationship to the second, is the moving and dynamic question of "What does it do?" Here belong all the traditional *physiologies* (*physio,* by convention, functions or processes), and the studies in behavior coming to be known by the term *ethology*. Here also belongs the knowledge of *Growth* (the monotheistic God of the economists), and its corollary *Death* (our consciousness of which is probably responsible for more intellectual inanity than any other item of our awareness).

Physiology is always d y n a m i c, whereas morphology is s t a t i c. I am reasonably sure that the personalities of the respective scientists are equally different. Physiologists differ from morphologists in dress, mannerisms, and certainly in the appearance of their offices and laboratories.

A *fourth* fundamental question is *"Where* is it?" Some people are basically, fundamentally, and overwhelmingly interested in *spatial* matters. Once they find out where a thing is, their curiosity is satisfied. Here are the chorologists and the *geographers (choro,* space). Unlike the two preceding specialties, plant and animal geographers may be found either in geography departments (where they are dominantly geographers—organisms are simply something to place) or they are in biology departments (where they are dominantly biologists—secondarily to be placed).

A *fifth* question, and bearing a pair-relationship to the fourth, is the compelling query of *"When* was it?", or *"When* will it be?" ,in short, the *temporal* concerns of mankind. Here are the chronologists and the *historians (chrono,* time), students of the past, as well as seers, prophets, Weather Bureau prognosticators, and others who have that human compulsion to try to forecast the future. As with geography, we find these people both in history departments (where they can be very poor biologists but where some of the finest chrono-biological research can be accomplished) and in biology departments (where they can be very poor historians, and where prediction—even if stated in the abstruse and elegant mathematical language of systems analysis—can be merely the orderly extension of a few quantified parameters, an extension which a good historian knows does not occur.)

The *sixth* and last question, in terms of this hexagonal pyramid, bears a pair-relationship to the first, which asked what is "inside." This time, the scientist asks "What is *outside?"* "What is the *environment* of the subject matter?" This use of the word ecology, as a point of view towards a subject matter (rather than as a subject matter itself) has been common on the European continent, and has much to recommend it etymologically *(oikos,* home, surroundings). (In America, however, the word ecology is linked with the entire subject matter of the 7th Level of Integration, biocommunities, and as such has been a course taught in biology departments. I am not inferring which usage is "right"; I am simply commenting on what is done.)

From the standpoint of our modelized interpretation, we note that

the components of the Subject Matter (first question) can become
the wholes of a next lower Level of Integration. For example; a bio-
community of plants and animals (Level 7), is composed of populations of various kinds or organisms (first viewpoint). But each such
population can well serve as a Subject Matter (Level 6) for a complete scientific study of its own. Also it is time to point out that the
Subject-matter-*plus*-its-environment (e.g., a bio-community at Level
7, plus its environment of soil and climate) becomes a whole of a
higher Level of Integration (an ecosystem at Level 8). Man-plus-his-
total-environment is the studiable whole of Human Ecosystem Science
(Level 9). Although the complexities of nature are by no means as
simply expressed as this sentence would indicate, the idea lends understanding to many phenomena that we observe in the world about us.

We have left the base of the pyramid to the last. It represents the
Seventh and last question, "How is it classified?" Classification, organization, systematization, *taxonomy* (the word was first used in botany)
is absolutely fundamental to each and every science. Even before the
current knowledge explosion, scientific data were far too numerous
for any person to remember bit by bit. Instead, we group and regroup,
arrange, and classify. Thus, the philosophy of taxonomy can be based
primarily on *any one of the six* faces contiguous to the base. Such
numerous potentialities are particularly apparent in the study of
"Vegetation" (the complex of plant-communities in the landscape).
Vegetation types may be classified primarily by their flora (composition), or by their physiognomy, or by their successional status (Vegetation physiology), or by similarities in spatial or temporal distribution, or by their environmental relationships, or any combination of
these. On the other hand, traditional "plant taxonomy" had become
stereotyped as a classification of the elements of the flora in terms of
their supposed evolutionary relationships, even though foresters, horti-
culturists and others used many other classifications of plants. Taxonomists are among the most orderly of scientists, with a name and place
for everything. They are the scientific analogues of the perfect house-
keeper, and the perfect stockroom manager.

There is one supplementary thought in regard to the use of this
pyramidal model of the Points of View. The model is only an approxi-
mation to nature, useful within limits for understanding the complexities of nature. It is not to be thought that the six sides of the pyramid
always remain distinct, even though the personalities of scientists may
tend that way. The interfaces between two sides may themselves
blossom forth as pregnant fields of investigation. For example, we
have physiological morphology, ecological anatomy, historical geography. When there is such a need, it will appear; but even in appear-

ing, its origin in the prime faces will be evident.

We have now constructed a geometric 2-dimensional model as an empirical classification of the existing sciences. This model can be thought of as a 9 x 7 chart, with the nine Levels of Integration as the horizontal abscissa, and the seven Points of View as the vertical ordinate. As an additional model and concept, each vertical column (the 7 Points of View of any one Level of Integration) can also be thought of as a three-dimensional hexagonal pyramid, with the base as the seventh (taxonomy) point of view.

We also find it desirable to recognize a kind of equivalence between the Compositional (inside) Viewpoint of a particular Level of Integration, and the entire vertical column of a *lower* Level of Integration. Somewhat similarly, we recognize a kind of equivalence between the Ecologic (outside) Viewpoint of a particular Level of Integration (the interrelationships with its environment, of a Subject Matter), and the material of a *higher* Level of Integration. This "diagonal phenomenon" is a corollary of the fact that parts of wholes can also be considered wholes in their own right, just as hydrogen and oxygen are not only wholes by themselves, but also parts of the water molecule.

Any such classificatory model as this, with a limited number of pigeon-holes and of little boxes in which to fit things, is bound to stimulate in some otherwise intelligent people what I call the Boundary Bugaboo. This kind of orderly person, seriously considering the model, will tell you he will accept it *if* you can convince him just how you can "define the boundaries" of each such subdivision. Clearly of course, nature does not approve of fence-like boundaries (these are wishful artifacts of our minds), but only of relative continuities between relative discontinuities. Our "fences" simply help us to understand the complexities, and we should never be fooled into thinking them anything else. I strongly suspect that those objectors who become upset about the Boundary Bugaboo are simply trying to project upon the rest of nature their own anthropomorphic separatism (as in an organism—once the umbilical cord is cut, there is a clearcut boundary between oneself and everyone else). Some day I shall catch such a bugabooist in bed with his wife, while she is casually chewing peanuts, and then I will ask him to define holistic boundary lines.

I have often presented the 9 x 7 and the pyramidal models in lectures and conferences, but never did I receive such a sober example of boat-missing as at a rather prestigeful meeting of academic abbots who were actually gathered together to consider this very problem. The geometric model was adjudged, not to open the way to new creative and innovative thinking, and to Ninth Level Integration thru all the departments of a college campus. To the contrary it was damned as not being forward-looking, but actually backward-looking, doing no more than to propose a rigid pseudo-classical system of cubby-holes more worthy of the Renaissance than of what we need today. I found this "Renaissance Reaction"—which should be called the Rice Effect, for reasons best known to myself and those there—far more revealing than my critic will ever know. If this is "Renaissance thinking," then this is precisely what I wanted to stress *did* occur, not only in my critic's own mind, but on his campus, and indeed in the entire academic world. Moreover, I see no great insult in saying that this is the way academics *do* think. The model does not seek to reveal reality in some new illuminating revelation of nature. It merely aims empirically to describe the way the little specialist-minds of academe have actually built mansions of many isolated rooms, with no connecting doors or passages. Until this architecture of academe is admitted, then accused, then accursed, we cannot even start to begin to commence to train the youth of our nation in the direction of integrative Ninth Level thinking involving the unity of man-plus-his-total-environment. The greatest hurdle in attaining any such goal are the professors of academe themselves. They are professors because they *are* specialists, specialists not by training but by original aptitude. The generalists—victims of this same Renaissance, for with the explosion of knowledge, the generalist was not only unneeded in society, he was left to wither on the vine as a crabbed crank—are the ones to lead us. They will not be found in academe, except by accident.

*   *   *   *   *

I guess it can be said that I am optimistic that such an integrated Ninth Level academic program can and will arise—or I would not be writing this book. On the other hand, I am under no illusion. It will not happen until college presidents are total-ecologists, or total-ecologists are college presidents. It will not happen until some campus, however small, has not only financial freedom but academic freedom, and by academic freedom I mean something far different than the freedom to be a dissenting activist to the detriment of everyone but oneself. It will not happen until a faculty, however small, can be found who are primarily generalists, not specialists, even if it means dropping some eminent Nobel Laureates.

I am also pessimistic. There are times I strongly believe that an all-embracing ecosystem science, fathering an ecosystem technology, in both research and teaching, is not destined to arise in our youthful and vaunted West (where there is a close race between technology and the archeologists of the future), but in the maturer climate of the ancient East. Perhaps this idea is rooted in two spheres of inquiry: comparative theology, and comparative linguistics, neither of which can be discussed adequately in these pages. We in the West are still immersed in an adolescent Judeo-Christian heritage, despite cries from cradles that "God is dead," and despite the attempts to raise St. Francis to the status of a saintly ecologist. Our Western heritage pivots on the assumption (as do all primitive cosmologies) that one's own tribe is God's Chosen, that the first man of the tribe was so created, that all other tribes are barbarians. Then the rest of nature was created to serve that man, and to be exploited by him. Our basic philosophy is "nature *for* man," and "man *against* nature." It is an immature, arrogant, conceited—and very ignorant—philosophy. This idea of "domination" and "subjugation"—the Biblical injunction of man's relation to nature—must be expunged from our thinking. Supporting such arrogance, and perhaps even an expression of it, is the indicative linguistics already discussed (p. 5-7)—one of the world's few—involving nouns, verbs, and modifiers, subjects and predicates, tenses and moods which—even if coincidentally—amply serves to describe *and to manipulate* (technology again!) the environment about us.

In theology, the Orient, older and wiser—even if it had to learn the hard way, by experience—knows and accepts the confines of its spaceship. Famines, floods and the limitations of nature are recognized and borne with patience, and a wisdom we do not have. Man there lives and operates *with* nature, as a part of nature—unless he has become a Christian technologist, running for a Yankee dollar, which he will soon spend, to be poorer than before. I look to Hinduism, Buddhism, and Taoism, to the spirit of Buddha, Confucius and Lao-Tze as the womb from which a humanitarian-oriented Human Ecosystem Ecology may yet arise.

> *Has God, thou fool! work'd solely for thy good,*
> *Thy joy, thy pastime, thy attire, thy food?*
> —ALEXANDER POPE (1688-1744)

*   *   *   *   *

The paragraph above was written in 1968. Yet in the dawn of 1970 the sun—symbol of a Total Environment—is rising in the West upon man in a landscape that may prove me superbly wrong. I refer to no movement in academe, in science, even in ecology, nor to any in ecumenical religion. These are still failing the needs of the times. I do refer to a situation within our youth that thereby may extend to embrace all mankind together with his environment, involving principles that subsume the lofty conceptions of the great religious leaders of all places and times, who have periodically risen to point the way to a higher evolution.

> *Those who would carry on great public schemes must be*
> *proof against the most fatiguing delays, the most morti-*
> *fying disappointments, the most shocking insults, and*
> *worst of all, the presumptuous judgment of the ignorant*
> *upon their designs.*
> —EDMUND BURKE (1729-1797)

AN ANSWER TO THE NEED: THE NINTH LEVEL*

by Harry E. Van Deusen

Scientists have a narrow soul, a short sight, and generally an underde-veloped heart. They are dry and inhuman, and so often ugly. All that be-cause they are burrowing without looking at any sky. I dream to open this sky, right in the line of their tunnels. [Written 1938]
—PIERRE TEILHARD DE CHARDIN (1881-1955)
Letters to Two Friends. 1926-1952

To the eye of the exact sciences, one of the most remarkable char-acteristics of life is its "additive" quality. Life propagates itself by ceaselessly adding what it successively acquires—like a memory, as has often been said. Every living being passes on to his successor the being he himself inherited, not merely diversified but accentuated in a given direction, according to the line to which he belongs. And all the lines, whatever their nature, seem in varying degrees and each after its own formula to move a greater or lesser distance in the general direction of greater spontaneity and consciousness. Something passes, something grows, through the long chain of living creatures. This is the great fact, or the great law, the discovery of which has transformed our vision of the Universe during two centuries.

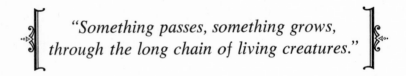

"Something passes, something grows, through the long chain of living creatures."

If we accept this fact and we are in basic agreement with the idea that life has attained through man the highest degree yet of inventive choice in the individual and social organization in the world commun-ity, then it is time, in fact absolutely imperative, that we begin to rethink our concepts of science, its opportunities, its duties, and yes, its restrictions. Science and its practitioners must grasp the facts that take place in all of us while we grow up. We become aware of our

---

*It is planned to elaborate the material of this section in another volume, the third of a trilogy, of which the present volume is the first.

family past, our present responsibilities, our ambitions and our loves. And this is nothing but a brief recapitulation of a far vaster and slower process through which the whole human race must pass in its growth from infancy to maturity.

Therefore, if science is to serve God and man towards full "maturity" and the ultimate dream of Teilhard's "Divine Milieu," science must "grow up"; it must shake itself out of its own discipline boxes; it must accept its social responsibility; it must learn truly to deal scientifically with the *total* living experience of man, by supplementing the knowledge of things in the environment and of the body machine with a science of human life. Only then will science be capable of a role in giving a larger scope to human potential freedom by providing a rational basis for option and action. We should no longer be satisfied with studying the biological machine whose body, brain, and environment can be altered and controlled by drugs and mechanical gadgets. We must become vitally concerned about the nature and purpose of man. This calls for a new science, a science based on far broader concepts than the scientists themselves have held in the past. Perhaps broader than we may be able to think.

*   *   *   *   *

To this end, to answer this need, I would propose a series of Regional Centers for the Study of Man and His Total Environment. The overall goals of these Centers would be to develop an integrated, ecologically oriented program to determine how man can live in productive harmony with his environment in an age of rapidly growing populations, environmental contamination, and accelerating technology and industrial development.

The interactions between man and his environment are now changing so profoundly as to signal the start of a new era in the human habitation of the earth.

Rapid population growth and accelerating industrial development have combined to alter the environment on a scale unimagined a few years ago. Water and air can no longer be regarded as infinite diluents of the waste of modern civilization; the impact of human activities on the *total* environment is now clearly perceptible and must be understood and controlled if human society is to survive and flourish.

The program of these proposed *Centers for the Study of Man and His Total Environment* should be oriented toward the ultimate goal of developing ecological theory concerning the establishment of homeostatic relationships between man and his environment.

The advancement of knowledge toward this goal will require a total systems approach. This total approach must proceed from the viewpoints of the component living organisms and nonliving features of the system, the structural arrangement of the parts, the distribution of the system in time and space, environmental relationships, and classi-

fication in relation to other systems, always developed around the
central theme of *total* environment.

Classical investigations of subsidiary ecological systems with reference to flora, fauna, microorganisms, climate, soils, hydrology, vegetation, and animal life are fundamental in this total systems approach.
More sophisticated studies involving the functioning of ecosystems with
regard to the dynamics of populations, productivity, the cycling of
mineral nutrients and contaminants, and the interactions of populations in numerical homeostasis should become a part of a total program. Since man is an integral part of total ecological systems, the
behavior of society itself in relation to environment is as relevant to
theory and understanding of the regulatory control of ecological systems as any of the non-human aspects.

These proposed Centers should each operate in a local, regional
"core area." Within these areas ecosystemic problems can be studied
using man and his relationship to his environment as functional wholes
in nature. A "heartland" of these core areas should be a series of
Natural Areas for basic biological research. The surrounding core
area of perhaps a 30-40 mile radius would form an area for the studies
essential to understanding the impact of human society on environ-
ments. Using the Centers as models to develop methodologies for the
study of total systems, the ecosystemic studies conducted therein could
have significance to the rest of the world as a whole.

The unique approach of the Centers is in studying man-and-his-
environment as a single organized whole, including *sociological, economic* and *humanistic* (cultural and artistic) problems as well as projects in the natural sciences, always using the ecological viewpoint as
a unifying biosocial mechanism at the Ninth Level.

Basically these Centers could be developed along five general lines
of activity.

1. To make available a critical landscape as a physical land base for
continuing long-term programs of research in basic ecosystem science.
This part of the program would pertain to Natural Areas in the core
area set aside under perpetual care and assured protection, as well as
open space used primarily for agriculture, forestry, wildlife, and recreation.

2. To promote and encourage a sound program of ecosystem science
research by offering Research Fellowships and cosponsoring research
proposed by educational and research institutions.

3. To sponsor seminars and symposia on the advances, the problems,
and the tasks ahead in the fields of a Ninth Level ecosystem science,
with full recognition of the humanistic artistic and ethical human
values.

4. To evolve interinstitutional multidisciplinary programs of liberal

education oriented toward understanding the structuring and functioning of the various environmental complexes. These programs would allow students to develop a deeper appreciation of human values and a better concept of man's place in the universe.

5. To develop an effective program in communications that will reach all parts of society and that will function as a "clearing house" for environmental information significant to the particular core area.

The Centers should provide environmental information and guidance to governments, schools and citizen groups concerned with total environmental problems. Methods and media that could be used by these centers would include newspapers, magazines, radio, and educational television. These Centers would also publish a series of Ecological Study Papers and Occasional Papers of a technical nature based upon the research carried out at each Center.

The program in research and education should function primarily at the graduate and undergraduate levels, even tho primary and secondary schools would not be neglected. Particular emphasis would be at the undergraduate college level in the belief that irrespective of professional goals, students may use this ecosystem oriented program toward attaining homeostatic adjustments of society to environmental resources.

*   *   *   *   *

The Centers I am proposing look to the future. For a century now, scientific activity has become both quantitative (in the numbers of individuals engaged in it), and qualitative (in the importance of the results obtained). It is a major—if not the principal—form of reflective activity on earth today.

That activity means that there must be some extremely powerful motive force to maintain and accelerate such a movement. Let us see if we can find an answer to the question.

In the words of Teilhard de Chardin, "What initially makes man a "scientist' (and this runs on from what seems is already present in the higher animals), would appear to be the speculative attraction of *curiosity* combined with the economic stimulus of an *easier life*. To discover and invent for pleasure as well as for necessity—to improve the conditions in which one lives:—this twofold need of diversion and comfort may rightly be regarded as the original impulse behind research."

At the same time we have to recognize that, accompanying the latest developments of knowledge, a new and much more powerful psychical stimulus is making itself felt in today's seeker: not simply the appetite of well-being, but the sacred and impassioned hope of attaining *fuller-being*.

If ecology—ninth level science—is taken to its extreme limit in a certain direction, can it effect our emergence into the transcendent?

To that question, I believe, we must answer that it can, and for the following reasons.

Although we too often forget, what we call evolution develops only in virtue of a certain internal preference for survival (or, if you prefer to put it so, for self-survival) which in man takes on a markedly psychic appearance, in the form of a *zest for life*. Ultimately, it is that and that alone which underlies and supports the whole complex of all the bio-physical energies whose operation, acting experimentally, conditions anthropogenesis.

In view of that fact, what would happen if one day we should see that the universe is so hermetically closed in upon itself that there is no possible way of our emerging from it—either because we are forced indefinitely to go round and round inside it, or (which comes to the same thing), because we are doomed to a real and total death? Immediately and without further ado, I believe, we would lose the heart to act, to live, and man's impetus would be radically checked and deflated forever, by this fundamental discouragement and loss of zest.

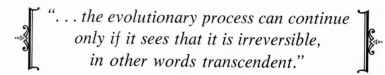

"*. . . the evolutionary process can continue only if it sees that it is irreversible, in other words transcendent.*"

This view can mean only one thing: that by becoming reflective the evolutionary process can continue only if it sees that it is irreversible, in other words transcendent. The complete irreversibility of a physical magnitude—inasmuch as it implies escape from the conditions productive of dis-integration which are proper to the physical world of time and space—is simply the ninth-level biological expression of transcendence.

Evolution, the way out towards a more glorious future that escapes total death, this is the hand of the God of today, gathering us unto himself.

INDEX

Bold face type indicates a major reference to the subject.

143

THE AUTHOR

Frank E. Egler is a professional ecologist of unusual parts. The eminent LaMont C. Cole said he was a man "with a social conscience, and also a logician and philosopher who is uncommonly intolerant of ignorance and mediocrity. He is no diplomat, and can be a devastating critic." This statement was published after the Entomological Society of America publicly censured the American Institute of Biological Sciences for printing a critical article on pesticides by Dr. Egler in their journal BioScience.

Dr. Egler was born in New York city in 1911. His formal education was a forerunner of things to come. It involved three preparatory schools, and eight universities stretching from Honolulu to Paris. In the course of these years, he garnered three degrees and various certificates. Characteristically, he claims he is spending the rest of his life trying to forget the "facts" then learned, while he has the highest admiration and esteem for a few great teachers he was privileged to know.

Tho he seemed destined for a conventional academic career—he became a professor at the age of 26—he soon dropped out of the confinement. Researcher, lecturer, author, educator, newspaper columnist, world traveler, his activities have included such other diversities as being a professor of physics, a writer in electrical engineering, seminar leader, and the handling of financial investment accounts, with sparring episodes in conservation, law and theology. Professional publications under his own name in ecology and conservation total over 150, in journals here and abroad. His affiliations, in addition to those at the State University of New York (Syracuse, forestry), Yale University, Wesleyan University, University of Connecticut include connections with the American Museum of Natural History, the New York Botanical Garden, the U. S. Forest Service, the Guggenheim Foundation (Fellow, 1956-58). There was an effort to initiate an experiment station for the chewing gum industry in Central America, and another effort with several industries to establish a sound management of Vegetation on the roadsides and rightofways of the nation.

Since the end of 1945, Dr. Egler has been based at his own property in the Berkshires, used for his field research. His writing continues. He is a "free-lance" to some, "retired" to others; while a few of his adversaries wish he really would retire.